Reflections
of God's Grace in Grief

Faythelma Bechtel

REFLECTIONS OF GOD'S GRACE IN GRIEF

Copyright © 2012 by Vision Publishers
Copyright © 2009 Faythelma Bechtel

All rights reserved. For permission to quote from this book, contact Vision Publishers. No permission is needed for 50 words or less.

ISBN-10: 1-932676-23-6
ISBN-13: 978-1-932676-23-5

Also Available as Ebook
ePUB-10: 1-932676-55-4
ePUB-13: 978-1-932676-55-6

ePDF–10: 1-932676-56-2
ePDF-13: 978-1-932676-56-3

First edition: 2009
Second edition: 2012

Printed in the United States of America

All Scriptures are taken from the King James Version unless otherwise indicated.

Layout and Cover Design: Lanette Steiner

For special discounts on bulk purchases, please contact:
Vision Publishers orders at 877.488.0901

For information or comments, contact:
Vision Publishers
P.O. Box 190 • Harrisonburg, VA 22803
Phone: 877-488-0901
Fax: 540-437-1969
E-mail: orders@vision-publishers.com
www.vision-publishers.com
(see order form in back)

Holmes Printing Solutions
8757 County Road 77 • Fredericksburg, Ohio 44627
888.473.6870

This book is dedicated to:

with love by

Come unto me, all ye that labour
and are heavy laden, and I will give you rest.
Matthew 11:28

Introduction

Dear Fellow Traveler,

These devotional meditations are born from the tears, sorrow, and pain of numerous trips through the valley of the shadow of death. These ventures with God began in 2004.

Looking back over the years seems like a mysterious nightmare, yet the truth of God's compassion and care have been a reality.

After caring for my mother for seven years in our home, she passed away on July 12, 2004. She struggled all the years of her life with enormous emotional swings. Today her disorder would be called bipolar. Caring for her was a colossal challenge and left me emotionally frazzled.

On February 14, 2005, our oldest daughter, Cynthia Joy Bechtel Kropf, was diagnosed with inflammatory cancer. It is a rare (only 1% of all breast cancers is inflammatory), fast growing cancer.

In June 2005, Cynthia wrote in an email, "Jonathan (her husband) and I still feel God's direction and incredible peace about the choice (their choice of treatment). Honest, I never dreamed it could feel this right.

"No, I do not feel I will come home and never have another problem or that I will never struggle with cancer again. Whether I live or die, my life is the Lord's. Yes, I could die tomorrow here or come home and die the next week. My life has been spared way beyond the time I should have lived from man's point of medicine already. I am, as are all of you, a miracle in God's hand. I just wish to convey to you the total restfulness and peace we have had through this experience, and it is our desire to watch and wait for God's further direction, which I am certain will be glorious."

On September 12, 2005, Cynthia passed away, leaving four children ages 12-18 and her devoted husband. She was buried on September 18, 2005, which would have been her 43rd birthday.

In August 2005, a month before our daughter's death, my husband was diagnosed with front temporal dementia. The doctor guessed that it likely began at least 10 years earlier.

Things in our lives had been strange and abnormal for so long, I could not think back to when it might have all begun.

The doctor explained that he would regress into childhood and then become like an infant. He said with Wilmer's strong, healthy body and at his age (65), he could be an invalid for 20 years—depending on how long he might live. This was not the news anyone wanted to hear.

Wilmer's health and behavior deteriorated much more quickly than the doctor speculated. Did that mean the disease had started at a much younger age, or was the mercy of God at work? No one wants to see a loved one lying as a vegetable for years!

My feelings of frustration, anger, depression, and confusion could fill another book, but one cannot grow while feeding on the negative—therefore, no book! Life became a perpetual hassle and trauma. Diapers, messes, eating garbage, running off, getting into everything that was not locked up, and disruption were the order of the day.

Working through rejection, the crushed dreams, and then grief of losing our daughter and my husband were almost more than my body and mind could handle. I was carrying a load I could not bear.

My cry was, "Why, Lord, why can't I handle this? You said You would not give me more than I could bear. What is wrong with me?"

In 2007, I finally had to admit I could no longer do what I felt was my duty to do. We moved Wilmer to a care facility. Only God knows the guilt, pain, and shame that is felt when you cannot do what you think you should be doing.

He seemed to adjust well and was always all smiles when I would go to see him. That was a special blessing with his kind of dementia—he always recognized his family and people he had known before his disease.

On April 21, 2008, just two days past his 68th birthday, Wilmer passed away of pneumonia, the immediate cause, with dementia being the underlying cause of death.

On his birthday, many of our family went in to celebrate with him. He had been very sick and unresponsive, but when he awoke and saw us all around his bed, he smiled and seemed to enjoy all the attention. He was more alert and responsive with smiles and laughs than we had seen him for months. (He had stopped talking months earlier.) He enjoyed our singing to him.

When some of us went back on Sunday, he was again very responsive until the afternoon, then we could tell he was going downhill.

When I asked the nurse if this alertness and responsiveness was normal for dementia patients, she replied, "It is most unusual for one in his condition to rally like this. I'm sure he knows where he is going."

"Thank You, Lord, for giving him back to us for this brief time," was the prayer of thanks on all our lips.

Searching for comfort from God's Word to me, writing these

meditations, and reading poetry and quotes of other hurting hearts has definitely been part of my healing process. It is my prayer that you will find some words of encouragement and comfort during your difficult journey through whatever dark valley you may be traversing.

Since this book has been published, I have lost our youngest daughter, Sonya Renette Bechtel Miller. She passed away at La Monte, Missouri, May 12, 2010. She had intestinal cancer that rapidly spread to the liver. She would have had her 41st birthday on August 14, 2010.

Her passing has wrung from my heart another collection of devotional meditations. Where can you go with a broken heart but to the Lord and His Word! I do not know if or when the book written in her memory will be printed, but the writing process is working its healing into my heart. She was the light in my dark window and now she is gone, yet God gives another light.

"I will love the light
For it shows me the way,
Yet I will endure the darkness
For it shows me the stars."

A fellow traveler,
Faythelma Bechtel

IN LOVING MEMORY OF
My Mother, Our Daughter, and My Husband

SOMEWHERE OVER THE RAINBOW

Somewhere over the rainbow
God has planned
A most glorious mansion
Over in Glory land.

Somewhere over the rainbow
I'll meet you.
We will sing together
A song that's forever new.

Some day when God shall call me home,
I'll wake up where clouds are far behind me.
Where tears and troubles are no more.
Far away from sin I will soar.
That's where you'll find me.

Somewhere over the rainbow
I'll meet you.
We will sing together
A song that's forever new.

> **FATHER, IF THOU BE WILLING, REMOVE THIS CUP FROM ME:**
> nevertheless not my will, but thine, be done.
> *Luke 22:42*

Acceptance is the lighthouse on the shore,
casting rays of hope, guiding my tossing
ship from the sea of despair and crashing
waves of grief to the calm harbor of God's grace.
God's goal for my life can only be accomplished
as I accept what He sends.
It is normal to ride the swells and crash with the
surging waves of grief for a period of time.
But it is not normal to want to stay out
on the wild dashing sea,
because after a time my distressed ship
will be smashed to pieces by the storm.
Help me, dear Father, to seek that
lighthouse of acceptance and follow its
beams safely to shore.
As I grow in acceptance—getting closer to
the lighthouse—the brighter the light
and the smaller the waves of grief will become.

Acceptance of God's will means the relinquishing of my own—the only safe way to reach the shore and God's goal for my life. I must even give up the "whys" and "if onlys" if I desire a peaceful landing.

In Loving Memory of Cynthia Bechtel Kropf

Three Years

Three long years of missing my sister dear,
Three long years of shedding many a tear;
Three years of wishing for a good long chat,
Three years of crying, "How I miss this or that!"
Yes, three long years have come and gone
Since my dear sis went to her new home.
Yes, it's been three long years
  Filled with so many mixed tears,
So many wonderings,
  So many ponderings.......
        Always thinking of sis......
But it's also been three long years
Thinking of what a privilege is hers.
She got to be the first to go—
  Tho' she left us all here below—
  The first of the family
        Our dear Lord to see,
  The first to leave
  And to be sin-free.
She is the one that is blessed—
Blessed with never another earthly pain,
Blessed with eternal comfort—she gained;
Blessed with joy beyond compare,
Blessed with not a struggle to share,
Blessed with peace, blissful rest,
Blessed with none, NONE of our stress......
        Yes, she is the one who is blessed!
  And now our father joins her there
  In all the blessedness so fair,
  Where neither pain nor sorrow they know;
  New bodies and minds God does bestow.
No, we can't wish them back,
We never would......
  But we miss them so!
  And God knew we would!

—Sonya Bechtel Miller

The Waves of Grief

The waves of grief dash over my soul;
They flood in again, roll after roll.
They slash me, they smash me, they crash me!
"Where are You, Lord? The billows control!"

I feel myself crushed in their arms cold
'Til I think surely 'tis death's threshold;
I fall, I crawl, and to God I call,
"Save me, O Lord, and be my stronghold."

You lift me gently and guide my sight.
"Trust and acceptance will point you right."
He informs, conforms, and transforms me
As I struggle forward toward the light.

> YESTERDAY'S GONE, OH MASTER;
> And tomorrow may never be mine.
> Today's a painful disaster:
> Just hold me, one moment at a time.

Today is____? It doesn't matter what day of the week! It doesn't matter what day of the month! Time becomes languid and unimportant the day a loved one dies. Time seems to stop totally the day a loved one is laid to rest. Yet those dates are painfully seared into the hole in your heart and will be forever remembered with pain. All other time is irrelevant.

I am left in the black abyss of grief. It's supposed to be the valley of the shadow of death. But there is not even a shadow here in this timeless chasm of blackness.

Are You in this hopeless void with me, Lord? I don't *feel* Your presence, but I have knowledge of Your goodness, greatness, and love. Help me to believe You are somewhere, guiding me through this deep, deep darkness.

Help me, Lord, to remember—

> *Death is only a mile marker*
> *I need to pass,*
> *Not a grave in which to stay.*

> **OH THAT MY GRIEF WERE THOROUGHLY WEIGHED,**
> and my calamity laid in the balances together!
> *Job 6:2*

Job certainly had a good reason to think his grief surpassed the grief of anyone else. He had a good reason for desiring his grief to be weighed. Then he could prove to his accusing comforters the huge burden of his calamity.

Lord, maybe I'm too much like Job, wanting to show others my grief is very great. If there were such a thing as a "grief-o-meter," I could prove the depth of my grief!

Why do we humans always think we are hurting more than someone else? What value is there in that assumption?

Am I looking for sympathy?
  For understanding?
    For help?
      For mercy?
        For forgiveness?

Lord, I am grief-stricken and I need all of the above. But vain is the help and understanding of man, especially from one who has never experienced a major loss in his lifetime.

Help me, Lord, not to expect from man what only God can give in the fullest measure. And help me to remember there are others suffering as much or more than I am.

Never does man know the force that is in him
till some mighty grief has humanized the soul.
—F. W. Robertson

> LET THE DAY PERISH WHEREIN I WAS BORN, AND THE NIGHT in which it was said, there is a man child conceived.
> *Job 3:3*

Job was devastated! ALL that had once been received as a blessing from God was now taken away. One can only imagine all the feelings that were swirling within him. So many feelings he couldn't even identify them. So many feelings that he completely despaired of life itself.

Yes, Lord, I have been there. At my daughter's deathbed, I told her I should be in that bed and she out here by my side. Things were backward and upside down.

Feelings are swirling within me. Such a blob of feelings, I cannot separate one from the other. Feelings, oh God, that people tell me are not "Christian." I really think You look at them as "human" feelings. And I'm so thankful You remember my "dusty" make up.

Help me, Lord, to identify my feelings—label and confess each one to You. Then, please sift the chaff from the grain, the incorrect from the correct, and help me bring my thoughts into captivity—remembering thoughts create feelings.

Feelings come and go,
but principles are undisturbed and stand fast.
—Richter

> WHO IS WISE, AND HE SHALL UNDERSTAND THESE THINGS?
> prudent, and he shall know them?
> for the ways of the Lord are right.
> *Hosea 14:9*

The ways of the Lord are right? When He took our loved one away? When He allowed such a painful separation? When He left her family without a wife and a mother? When He left me with an empty, broken heart?

Was it His will that she die, or was her death the result of the curse of sin? Sin is the bearer of pain, sickness, and separation. We are living in a world that groans under the weight of sin's curse. Yet, God is over all and controller of all.

Lord, help me to be wise and prudent that I might understand that Your ways are RIGHT. You hold the blueprint for each life. You decide the final outcome according to that person's choices.

May I rest in the fact that You are sovereign and all knowing. Only by Your grace can I understand the rightness and justness of Your ways. You have a reason for everything You do and a schedule for when You do it.

> *Softly the leaves of memory fall,*
> *Sadly we stoop to pick them all.*
> *Unseen, unheard, she is always near,*
> *Still loved, still missed and very dear.*
> *Just as she was, she always will be,*
> *Treasured forever in our memory.*
> —Author Unknown

> COME YE YOURSELVES APART INTO A DESERT PLACE
> and rest a while.
> *Mark 6:31*

This verse says it is all right for me to give myself some time out. It's all right to take time to think, to take time to hurt, to take time to cry, to take time to heal. It's all right to take time away from people, away from work, away from pressure.

Actually, Jesus is calling me away to rest awhile in His loving arms. Oh, I need to see the light of love in His eyes. I need to be aware of His understanding. I need to feel the stability of the strength of His arms.

How I long to feel the compassion of His comfort, and to know the greatness of His grace during my dark hours.

I will come apart and rest awhile and learn more about Your unchanging character, oh Master. I will accept Your offer of comfort, of repose, and of security.

Though working through my grief is in no way restful, yet being away from the rush of life is restful. Knowing You really do care is restful.

To will what God doth will, is the only science that gives us rest.
—Longfellow

> **YEA, THOUGH I WALK THROUGH THE VALLEY**
> of the shadow of death, I will fear no evil, for thou art with me.
> *Psalm 23:4*

O Lord, this valley is so dark, so dreadful, and so painful.

   Are You really there?

      Do You really care?

         My faith is so small—I see no light at all.

But there is a shadow! A shadow means there is light somewhere.

When I cannot see the light of God's face, can I trust in the shadow of His wings? Can I trust that He is in this valley, walking with me, protecting me from Satan's dreadful darts of fear and hopelessness?

"There is a safe and secret place
beneath the wings divine...
Oh, make that refuge mine."

> *A bereaved person is like a duck: above the surface,*
> *looking composed and unruffled...*
> *below the surface paddling with all his might.*

> WHEN HE HATH TRIED ME, I SHALL COME FORTH AS GOLD.
> *Job 23:10*

I have sometimes been a people-pleaser. I don't want to feel pressured to handle my grief in a special way so as to please people—so they might see how strong I am, or how well I am dealing with my grief.

Lord, help me to forget people! Help me not to allow myself to be pressured or pushed or programmed to fit other people's mourning molds. This is my personal mourning. What people think of me during this time of sorrow is unimportant.

This pain is between You and me, dear God. How I deal with it is Your concern. You are trying me in Your mighty refining furnace, and the end results are the most important to You.

Oh, guide me through each day of this painful process,
so that the end will produce a vessel You can use.

> MY DAYS ARE SWIFTER THAN A WEAVER'S SHUTTLE,
> and are spent without hope.
> *Job 7:6*

Did Job in his terrible grief feel the same about time as I do? How can it be that time seems to fly by and creep at the same time?

Two weeks have passed since the funeral—two long weeks of living
   In a daze of grief,
      In a haze of reality,
         In a maze of painful thoughts!
It seems life can't go on, but it will.
It seems the world has stopped, but it hasn't.
It seems You are very far away, but You aren't.
It seems You have forgotten me, but You haven't.
Oh, Lord, remove this daze with Your love.
Remove this haze with Your light.
Remove this maze with Your wisdom.

All that we love deeply becomes a part of us.
—Helen Keller

> O REMEMBER THAT MY LIFE IS WIND:
> mine eye shall no more see good.
> *Job 7:7*

Was Job to the end of his rope? What more hope was there in life for him? He had lost ALL his children and all his wealth. He had lost his health. The vast majority of what comprised his life was gone.

Looking at the immensity of Job's pain and grief should make mine smaller. But no, that is not the way it works. Each person's grief is painful in a way that only he knows, and each person must work through his own grief alone.

A part of me is gone—a part that can never return. It is a part that I cannot define. The beauty of the flowers reminds me of something painfully missing. The falling leaves remind me of a missing part. The scampering squirrels and singing birds bring to my mind that something is tormentingly missing. A song, a fragrance, a ripple of laughter, all these tear at my heart with haunting memories. Lighting a candle or bringing in a bouquet ends in a flood of tears. Family gatherings burn and sting because of a missing part.

Lord, help me to learn to live a new
"normal" life with one part missing.

17

Feelings buried alive, never die.

Lord, why do I struggle with the reality of the finality of the obvious? You created us with an immense love of life. Now, Lord, I ask You to give me acquiescence to death.

Help me to accept this loss. It seems so unreal; but the horrid pain in my heart tells me it is true.

I need help.

   I need Your help, God.

      I need the help of those I love.

I can't, and I won't, stuff this pain into the bottom of my heart. I bring this pain to You, and I ask You to relieve it and, in time, to take it away. I know such a devastating pain can't disappear at once, but I am confident it will diminish with time and with the healing of Your touch.

I don't want this pain to haunt me in years to come. I want to deal with it now. You will have to teach me how.

The pattern of our lives has changed—
for time has brought sorrow.
The pattern must be rearranged to fit a new tomorrow.

> HEAR MY CRY, O GOD; ATTEND UNTO MY PRAYER.
> From the end of the earth will I cry unto thee, when my heart is overwhelmed; lead me to the rock that is higher than I.
> *Psalm 91:1, 2*

Suddenly, my mind is filled with the nightmare of death—the death of my beloved! Oh, God, hold my breaking heart. My heart is overwhelmed. Don't let it fly apart into a million pieces.

Please help me to remember, *Let not your heart be troubled . . . I go to prepare a place for you* (John 14: 1, 2). Oh, yes, my Sovereign God; You have prepared a place for my beloved. A place of rest and peace. A place of no more pain, sorrow, or tears.

May I let her rest in peace. Help me to remember, I must always keep my heart prepared that I might dwell in the place prepared for *me*.

Remove all bitterness and doubt from my mind. Fill me with confidence that You do all things well.

Please, Lord, help me to the top of the Rock that is higher than I!

Persevere through the difficult patches,
and better times are sure to come.

> COME UNTO ME ALL YE THAT LABOUR AND ARE
> heavy laden and I will give you rest.
> *Matthew 11:28*

Lord, I am weary and worn, teary and torn.

Each morn seems too heavy to be borne.

Yes, that is the way of my grief. It relentlessly tears my heart apart until I am exhausted.

Please, help me to redirect my emotional energies. Help me to invest my energies in helping other hurting people. Surely helping someone else can't be as tiring as crying for myself.

Help me to find the assignment You have for me today. Help me to lift someone else's load. Help me to adjust my focus from my own heartache to the pain of someone else.

Help me to remember, dear God, that nothing done *by* You can be a mistake or done in vain. Neither can anything done *for* You be a mistake or done in vain.

I will come to You in the morning, and You will give me quietness and rest as I trust in Your compassion to see me through this time of deep pain and frustration. I will walk in the light of Your love and share it with someone else.

Teach me to walk in the light of Your love
and share it with someone else today.

> PEACE I LEAVE WITH YOU, MY PEACE I GIVE UNTO YOU:
> not as the world giveth, give I unto you. Let not your heart be
> troubled, neither let it be afraid.
> *John 14:27*

The Passing Bell

It slowly tolls. It tolls, "Good-bye
To life, and youth, and hope, and spring!
The sum and end of everything
Is—man is born and man must die.

"Brief as the Summer is our life,
As sure as Winter our decay;
There must be night if there be day,
And there is rest if there be strife!

"This is the end of Spring and Youth,
Spring ends in Winter, Youth in death!"
Nay, with its every bud and breath,
God's world proclaims a nobler truth!

The breathing world of flowers and men
These answer back the passing bell:
"Old year! Old life! Farewell—farewell!
But Youth and Spring shall live again!"
—E. Nesbit, 1880

Be prepared for death; and death or life shall thereby be the sweeter.
—Shakespeare

> FOR MY THOUGHTS ARE NOT YOUR THOUGHTS,
> neither are my ways your ways, saith the LORD.
> For as the heavens are higher than the earth, so are my ways
> higher than your ways, and my thoughts than your thoughts.
> *Isaiah 55:8, 9*

When I ponder these verses, I have to stop and acknowledge, Yes, Lord! You hold the world in Your hands. You see each sparrow that falls. You know the number of the stars in heaven and the hairs on my head. You created me and gave me so much. You determined into which family I should be born. You gave us our family. You are the almighty God!

You also have the right to take away. It is Your privilege to decide when someone's days on earth should end. You know when to bless, and You know when to cause pain. You know when to bring trials, and You know how to aid us through our trials.

You are not accountable to me for the circumstance You bring into my life, but I am accountable to You for my attitudes and how I deal with what You send. Thank You, Lord, for being so much greater and knowing so much more than Your feeble creation.

Blessed are the trials that add sweet fruit to our lives.

> As the mountains are round about Jerusalem,
> so the Lord is round about his people (me).
> *Psalm 125:2*

I feel so different, so set apart from "normal" people. Because of my grief, I am not like others who have not yet experienced such grief. I think differently. I react differently. I perceive things differently. I am not myself. It is very difficult not to feel self-conscious when I am out with people. It is painful to be alone, and it is painful to be with others.

I'm tired of sticking out like a sore thumb, throbbing with pain and sorrow. Please, just let me fade back into normalcy. But what is "normal"? That is such an ambiguous term—interpreted differently by everyone.

There will be no "normal" life as it was before my loss. Lord, assist me in my adjustment to a "new normal" and help me to find comfort and strength in that "new normal." Just surround me with the mountains of Your love and mercy.

What is "normal?" Only my Father knows!
It may be a storm for me; but for you a rose.
For some, "normal" is fame and earthly gloss,
For others it is pain, sorrow, and loss.
When feeling "abnormal" think of God's love,
Prepare for the "new normal" in heav'n above.
—Faythelma Bechtel

> AND WHOSOEVER OF YOU WILL BE THE CHIEFEST,
> shall be servant of all.
> For even the Son of man came not to be ministered unto,
> but to minister, and to give his life a ransom for many.
> *Mark 10:44, 45*

While on earth, Jesus was always only *a Christ distance away* from those in need. He never passed by any hurting soul. He went out of His way to find the distressed and sorrowing.

Help me, Lord, to remember You still are only *a Christ distance away,* even in the valley of grief.

At times You seem so far away. Everyone seems far away. People avoid me; they pass by me. They're afraid to speak, lest they would say something wrong. They're afraid to stop by, lest they would have to deal with my crying.

Oh, Lord, where are Your servants who are willing to be only *a Christ distance away* from pain and suffering? Most people keep themselves miles and miles away, lost in their busy world.

After I make it through this dark, lonesome valley, oh God, help me always to be only *a Christ distance away* from other hurting souls. Teach me to weep with those who weep!

The rainbow is the smile of God, made up of tears and light.

> LORD, MAKE ME TO KNOW MINE END,
> and the measure of my days,
> what it is; that I may know how frail I am.
> *Psalm 39:4*

The brevity and frailty of life is distressing and painful. What might tomorrow bring? Many lives have been taken in a moment, snapped like a thread. Lord, when will my moment come?

Your counsel comes to me from Psalm 90:12.
So teach us [me] to number our [my] days,
that we [I] may apply our [my] heart[s] unto wisdom.

» Thank you for prompting me to remember to:
» Never anchor my hope on a life so frail and brief.
» Never set my heart on earthly treasures.
» Never leave unkind words to reconcile for tomorrow.
» Never fear the unknown when the all-knowing God is my guide.
» Never count on tomorrow to do what God prompted me to do today.
» Never forget, eternity is sure.

A wise man knows his own ignorance;
a fool thinks he knows everything.

—Simmons

> HEAR MY PRAYER, O LORD, AND GIVE EAR UNTO MY CRY;
> hold not thy peace at my tears.
> *Psalm 39:12*

Where are You, oh God? Do You hear the prayer of my tears? *I was* [am] *dumb with silence. My heart was* [is] *hot within me* (Psalm 39:2, 3). I have no words to offer You—only tears and more tears. Do You see? Do You hear? Do You know?

It has been said, "Tears appeal to the Divine pity in a special way."

To the widow of Nain, You said, "Weep not!" You gave back to her a living son. To Mary Magdalene, You said, "Why weepest thou?" Then You Yourself became her joy. When Lazarus died, You wept with his grieving sisters. Then You brought Lazarus back to life. You *offered up prayers and supplications with strong crying and tears* (Hebrews 5:7) that the heavenly Father might remove that bitter cup of suffering. But that cup was not removed, and You obediently went to the cross for me. This was the ultimate demonstration of Your love and care.

"Oh, yes, He cares, I know He cares. His heart is touched with my grief." By Your death and suffering, You have given my loved one eternal life, eternal health, and eternal joy.

Oh, may I focus on what You have given and
not on what You have taken.

> THE LORD IS GOOD UNTO THEM THAT WAIT FOR HIM,
> to the soul that seeketh him.
> *Lamentations 3:25*

Oh Lord, this valley of grief is so deep and dark and long. There is no end in sight. The fearful shadows of my feelings prod my every step. The shadow of death hangs over me like a phantom of fate.

My heart longs for one who is not. My eyes ache for the sight of her face. My ears throb for the sound of her voice. Oh God, the hole in my heart is so huge and painful. Will it never heal? Will my tears never stop? Will the pain never disappear? Will the shadows never fade?

Help me patiently to wait for You to accomplish Your perfect will in my life. There has to be a good reason for this loss, but I must accept the fact that I may never understand or see the reason. Help me to seek You daily, and in that seeking, I will come to know You better and realize Your goodness even in the pain of this loss.

Lord, help me recognize the blessings You give me,
and help me have a thankful heart.
I owe You much, far, far beyond what I can ever pay.

A Word From Your Loved One

There Is No Death
There is no death! Although we grieve
When beautiful, familiar forms
That we have learned to love are torn
From our embracing arms—
They are not dead! They have but passed
Beyond the mists that blind us here
Into the new and larger life
Of that serener sphere.
—J. L. McCreery, 1835-1906

Death is not, to the Christian, paying the debt of nature,
as it has often been called.
No, it is not paying a debt; it is rather like bringing a note
to a bank to obtain solid gold in exchange for it.
You bring a cumbrous body which is worth nothing,
and which you could not wish to retain long;
you lay it down and receive for it,
from the eternal treasures, liberty, victory,
knowledge, and rapture—
(a glorious new eternal you!)
—John Foster

> GIVE GRIEF WORDS; IT WILL HELP IT DEPART
> and heal your heart.

Lord, my grief has short-circuited my brain. I have an overload of pain. I can't think straight. I can't remember. I can't make decisions. I can't say the right things. I can't hear things correctly. I can't face people. I can't handle even the commonest situations. And yes, I was taught never to say "I can't" when I was in school. But school could not prepare me for dealing with losses in life.

"Grief does strange things to you," I've been told.

"What you are going through right now will not last," I've been promised.

"Slowly, your brain will return to normal functioning, and your memory will come back," I've heard from those who have walked this painful path before me.

Jesus promised, *I am with you always, even unto the end of the world.* "I can do all things through Christ which strengtheneth me."

Thank You, God. Things will get better.

GRIEF SHARED IS GRIEF IMPAIRED.

Grief has a way of fogging your thinking, embittering your mind, disabling your creativity, hindering your desire to go on with life, numbing your feelings toward others, driving you into hibernation—if you let it.

When you feel these dreadful things happening to you, find someone with whom you can share your grief. You have likely been spending too much time alone. Always remember, "Grief shared is grief impaired." Much of the punch, pain, and power of grief is reduced when shared.

But why is it so difficult to find someone with whom you can share? It seems no one enjoys the house of the mourner—especially those who have never experienced real grief. Yet, Lord, You sent Job his "miserable comforters." Even though their words didn't help him, they roused within him feelings that needed to be expressed and gotten rid of. Their accusations goaded him to work through his grief.

Lord, send me what I need to rid myself of all unhealthy feelings.

> THE LORD GIVES STRENGTH TO HIS PEOPLE;
> the LORD blesses his people with peace.
> *Psalm 29:11*

Lord, am I not one of Your people? Where is Your strength in my time of weakness? Where is Your peace in the turmoil of this loss?

Oh, I must think deeper—past the exhaustion and turmoil—past the pain of this hour.

Indeed, You have always lifted me after times of crushing sorrow. You have carried me on Your wings when I was faint. And, yes, deep down under the turmoil You have placed a quiet trickle of peace.

A peace that seems much unobserved during torrents of sorrow, but no less is there all the while. A peace in knowing that You make no mistakes. A peace in understanding that Your ways are not my ways. A peace in knowing that our dear departed one has no desire to return to this pain-torn earth.

May I every day be more aware of Your love and care. Help me to release my aching heart to Your strength and peace.

God harrows our souls, making us long for something we
cannot have, in order to reveal to us what He wants us
to have, which in the long run is far better.

> **STRENGTHEN YE THE WEAK HANDS,**
> and confirm the feeble knees.
> Say to them that are of a fearful heart, Be strong, fear not:
> behold, your God will come . . . he will come and save you.
> *Isaiah 35:3, 4*

A president of Harvard University once kept a turtle on his desk with this inscription written under it, "Consider the turtle. He makes progress only when he sticks his neck out." Have you ever seen a turtle move forward while his head was enclosed within his shell?

How like an immobile turtle we become when we're going through grief. It is so easy to crawl into our shell of sorrow and ignore the world around us. We fear sticking out our necks. We fear making progress. Our world has stopped, and our shell of pain encases us. What is there left to live for? We are drained of motivation and inspiration.

But as we cling to the Rock and cry out for strength and courage, we will find ourselves sticking out our necks more often. Faith in God will strengthen our weak hands and feeble knees.

Life is a risk, but is our neck the most important part of us? No, our soul and eternal destiny are most important.

My grip on God's grace will move me forward.

God never meant for my valley experience to be my grave;
He meant it to become a higher plateau of life.

> FOR THE ARROWS OF THE ALMIGHTY ARE WITHIN ME, the poison whereof drinketh up my spirit: the terrors of God do set themselves in array against me.
> *Job 6:4*

Have you ever felt like God was shooting poison arrows at you? Job did! His trials were so intense that he felt exhausted in spirit. Where do intense trials and an exhausted spirit leave you? They leave you looking for someone to blame for all your pain.

Yes, I'll have to admit I've done some blaming also. I've blamed the doctors, I've blamed circumstances, I've blamed myself, and I've blamed God.

The poison arrows came fast and furiously. I felt that God was picking on me, punishing me, and demanding more of me than I had to give. Did God think I was super-human? Didn't He realize that one human can take only so much?

Job had those same feelings. But he had no idea what was going on between Satan and God. Neither do I.

Oh, Lord, forgive my blaming heart!

> TO HIM THAT IS AFFLICTED PITY SHOULD BE SHEWED
> from his friend; but he forsaketh the fear of the Almighty.
> My brethren have dealt deceitfully as a brook,
> and as the stream of brooks they pass away.
> *Job 6:14, 15*

When I look to people for sympathy and understanding, I usually don't find it. Neither did Job. He looked to his comforters for compassion and empathy and received derision and blame. He goes on to say that, even if he had forsaken the Almighty, they should have shown him kindness.

Job compares their friendship to a brook that overflows in the winter when it is not needed and dries up in the summer when most necessary.

Most people who are busy with "normal" lives don't have time or mind to consider the pain, emptiness, or work load of someone who has suffered the loss of a loved one.

Don't be too shocked and disappointed to find that some of your friends become as this poet expressed it.

> *He is gone from the mountain,*
> *He is lost to the forest,*
> *Like a summer-dried fountain,*
> *When our need was the sorest!*

> HE WAS DESPISED AND REJECTED OF MEN;
> a man of sorrows, and acquainted with grief.
> *Isaiah 53:3*

It seems that there are those who believe that grieving is not a Christian happening. "Your loved one died, went to heaven, hallelujah, praise the Lord, there's nothing to be sad about," is their theme.

Oh God, I can't deny my pain, and I don't believe You want me to. Even Jesus was *a man of sorrows and acquainted with grief.* What caused Him sorrow, and what gave Him grief? Mankind!

His very own creation refused to hear His teachings and rejected His life of compassionate giving. He was even forsaken by His chosen ones.

He had an immeasurable capacity for love and a boundless capacity for sorrow, because He was a perfect Man and wholly God. The weight of the sins of the world was a crushing load.

Jesus knew life as a school of grief. He has called some of us humans to experience and learn in the same school. To some, the loss is large; to others, the loss is less. To all, He says, "Gird up thy loins like a man, and learn of Me."

Oh, Lord, keep my heart and mind open to Your lessons.

> GRIEF IS THE PRICE YOU PAY FOR THE ABILITY TO LOVE.
> –Joanne Petrie

My loss is like a huge wave on the sea of grief. It doesn't matter which direction I turn, the wave is always there waiting to heave me into its dark abyss. Then God reaches down and gently lifts me back up onto the wave's crest.

It's a continual up and down motion of emotions. Am I making progress? Looking back to those first painful weeks, oh, yes, there's been a bit of forward travel. But why so much pain? I loved and lost. Dare I love again?

Life with family relationships, church relationships, and social relationships is all about love. If I take my heart and hide it away from all possibility of ever getting hurt again, I will sustain an irreparable wound and aloneness. Never to love again is an option I do not want to take. God created me to love and to be loved.

It may be difficult to remember, but grief is a LOVE word. Grief is all about love. If I had no love, I would have no grief. Grief has a way of showing me my capacity to give and receive love. Grief is the price I pay for my ability to love.

It is a beautiful necessity of our nature to love.

> GREAT IS OUR LORD, AND OF GREAT POWER:
> his understanding is infinite.
> *Psalm 147:5*

G od's
R eprieve
I ntended for our
E motional
F ortification

Death (and numerous other losses) presents many emotional challenges.

In God's great love and understanding, He allows us to grieve, which provides us with a way to release the multitude of feelings that almost smother us.

Grief is a way to discharge and recharge emotionally, thus creating a fortress for our bleeding heart.

GRIEF—though a negative word in the minds of most of us, it is sent to produce positive results.

That's a fact, Lord; I can't comprehend it at the beginning of my journey, but help it to become more obvious as I come out of this dark valley.

Please let this valley experience make me:
better—not bitter,
  caring—not overbearing,
    assisting—not resisting,
      understanding—not demanding,
        perceptive—not deceptive.

Grief hallows hearts even while it ages heads.

—Bailey

> I AM WEARY WITH MY GROANING;
> all the night make I my bed to swim;
> O water my couch with my tears.
> *Psalm 6:6*

Someone has said, "When you lose a parent, you lose your past; when you lose a spouse, you lose your present; when you lose a child, you lose your future."

Lord, I've not only lost my past, present, and future, I've lost *me!* I am no longer myself with my familiar roles. I am no longer a daughter—except by historical fact. Now that my used-to-be loving husband is living in his demented world, I am no longer a wife. I am a caregiver, a financier, a housekeeper, laundress, transporter, and lawn and garden manager. I am no longer the active mother of my deceased daughter. I am an active grandmother trying to fill a tiny spot in a huge vacancy.

It is difficult to find my place in this new life. I'm like a "square peg in a round hole." I don't fit anywhere, Lord, except in the hollow of Your hand. Help that place to be enough for me.

Dear friend, is your real "me" feeling lost? There's room for you also in the hollow of the Lord's hand.

Hope Ever

The sun will shine and the clouds will lift,
The snow will melt though high it drift;
Across the ocean there is a shore,
Must we learn the lessons o'er and o'er?
To know there is sun when clouds drop low,
To believe in the violets under the snow,
To watch at the lows for the land that shall rise—
This is victory in disguise.

—Author Unknown

> *Job's friends:* SO THEY SAT DOWN WITH HIM UPON THE GROUND seven days and seven nights, and none spake a word unto him: for they saw that his grief was very great.
> *Job 2:13*

    Those first seven days were the wisest spent days in Job's friends' lives! They recognized Job's great grief and kept their mouths shut.

    Grief is not a subject to be debated; it is not a problem to be solved. It is simply a statement that you loved somebody and lost.

    Job likely suffered more losses than any of us ever will. There were no words, no solutions, no remedies for his pain.

    It might have been wise if Job's friends had turned and gone home after the seven days. But they stayed and tried to play God by determining the cause of Job's trials.

    Oh Lord, give me grace and a loving spirit when *miserable comforters* say things best left unsaid.

> *If God had no more mercy on us than we have*
> *on another, the world would have burned up long ago.*

> THAT HE WOULD GRANT YOU, ACCORDING TO THE RICHES of his glory, to be strengthened with might by his Spirit in the inner man.
> *Ephesians 3:16*

That's what I need, Lord, strength to bear what seems unbearable. Though I cannot feel comfortable in this situation You have brought into my life, please give me comfort.

I cannot find external security, not in my associations, not in my job, not in my tears. Lord, I must find the internal ability—that strength in the inner man.

God's comfort does not make me cheerful in my deepest distress, but it makes me competent—able to cope. God's comfort does not remove the trials, but it supplies the strength to work through them.

Help me, Lord, to remember Your aim is not to make and keep Your children comfortable, but conformable into Your image—something most often accomplished through painful trials and sorrowful distresses.

Oh, may I learn what You are teaching me during this time!

> TRUTH FOREVER ON THE SCAFFOLD,
> WRONG FOREVER ON THE THRONE,
> Yet that scaffold sways the future, and behind the dim unknown,
> Standeth God within the shadow, keeping watch above His own.
> —*James Russell Lowell*

With all the pain and evil in the world, do you ever wonder, *Is God really there? Does He really care?*

Yes, when my heart is pained too deeply for words or song, I have wondered, *Is a caring God really there?*

Yet, when my life collapses, my friends fail, and I feel as though I've fallen into a deep, dark well, on whom do I call? Who understands my distraught thoughts? Who feels the throbbing pain of my heart? Who gathers my tears and listens to my cries? ONLY Christ—*a man of sorrows and acquainted with grief.*

When I cry out to God, I affirm His ability to relate to my pain. I acknowledge His control. I admit He is wise and good in spite of the evil, sickness, and death in this world. I display a faith that, many times, I do not feel.

Remind me, Lord, that often the hours of grief, pain, and defeat reveal spiritual strength better than mountaintop experiences.

Who knows the balance between the soul's calamities and the heart's griefs? God alone!

> THE LORD IS RIGHTEOUS IN ALL HIS WAYS,
> and holy in all his works.
> *Psalm 145:17*

Is suffering and death caused by fate, nature, sin, or God? How I answer that question will make a great difference in how I deal with loss and grief.

The curse of sin causes disease, sickness, and death. Earthquakes, tornadoes, storms, and floods cause sickness and death. What does the Creator of life have to say about sickness and death? *All power is given unto me in heaven and in earth* (Matthew 28:18). Surely a God of ALL power has control in all situations of life and death. My inability to understand God's decisions reveals my humanity and exhibits the magnitude of God's wisdom.

Just because the death of my loved one has caused me endless pain, does not mean that God is bad, or that He made a mistake. Though God's purpose is hidden from me now, and perhaps until eternity (and then it will not matter), does not mean God made a heartless blunder. God is never heartless, and He never blunders. Help me, Lord, to rest in Your omniscience.

We should give God the same place
in our hearts as He holds in the universe.

> BLESSED AND ENVIABLY HAPPY,
> *[with a happiness produced by experience of God's favor and especially conditioned by the revelation of His matchless grace]*
> are those who mourn, for they shall be comforted.
> Matthew 5:4, Amplified

Have you ever thought about envying someone who is dealing with grief? I most certainly never have. Yet this verse implies the grieving one is experiencing God's grace in a way that is to be envied.

James Means said, "Not to be pressed above natural ability is never to know the sustaining grace of an omnipotent God. Not to know that sustaining grace in dire circumstances is to miss the greatness of his joy."

What presses you above and beyond your natural ability more than a tragic loss? That loss may be a loved one's life, a loved one's mind, or a loved one's ability to care for himself. Or it may be a more personal loss, perhaps your own inability to care for yourself, or no one to love and care for you. There are tragic losses too numerous to list.

When my loss has proven my powerlessness, and I reach out for God's help, then God's power will display His limitless grace in my life.

Thank You, Lord, for Your comfort!

Grace is abundant—more than I need.
Grace is universal—available for all.

> IT IS ENOUGH; NOW, O LORD, TAKE AWAY MY LIFE;
> for I am not better than my fathers.
> *1 Kings 19:4*

Thank You, Lord, for not becoming angry with us when we unwisely speak our minds in the midst of our grief. Thank You also for sometimes not giving us what we ask for.

Grief has a way of making you feel you've had enough and it would be easier and better to die than to live.

Often a loss causes you to lose your sense of direction and purpose. Everything gets out of focus. The Lord adjusted Elijah's focus by giving him rest, food, and a mission. Frequently, we need the same adjustment when we're going through grief.

Whenever I feel ready to succumb to that useless, hopeless feeling, I quickly find someone I can reach out to, or I go to my sewing machine and sew clothing for children in third-world countries.

There is something satisfying and rejuvenating about finding those with a need and helping them. Suddenly, you realize you are not the only one hurting and in need of a little love.

*There is no exercise better for the heart
than reaching down and lifting people up.*

> HAVE MERCY UPON ME, O LORD;
> for I am weak: O LORD,
> heal me: for my bones are vexed.
> *Psalm 6:2*

Indeed, the Psalmist's mental distress has produced weariness and bodily ailments. He sounds just like I feel in my grief! And yet, there are people who urge me to "get over it!"

Grief is not a problem to solve or to get over. Grief is not a disease, like measles, that you will recover from in a couple weeks. Grief is more like an amputation than a temporary sickness.

It takes an amputee weeks, sometimes months, to stop feeling the limb that isn't there. Phantom pain often causes him to try to use the missing limb. In time, most people will adjust to their loss, but the reminder of their loss is ALWAYS there.

So it is with grief—it may take months, sometimes several years, to adjust to life without a loved one, but you will adjust as you learn to accept God's difficult plan for you. Yet there will always be so many reminders of what you have lost.

Strength is born in the deep silence of
long-suffering hearts; not amidst joy.

Only God's World

Why trust this God who labeled His world good,
   With perfect seasons carefully designed,
     If senseless accidents can still intrude
   And rend the closest ties of humankind?

What world but God's endures loss and survives,
   Can bear and beautify, can make grief seem
     The awful tragedy it is, in that our lives
   Require divine involvement to redeem?

For in a world that claims no God but chance,
   There chaos is the norm and trust deceived.
   All grief's a joke where all is happenstance,
   All love a waste where none can be believed.

If you would have your sorrow honored, keep
Your faith in God Who sits with you to weep.
     —E. Gingerich, Ontario, used by permission.

Patience is an excellent remedy for grief,
but submission to the hand of Him Who
allows it is far better.

> THE ETERNAL GOD IS THY REFUGE, AND UNDERNEATH
> are the everlasting arms.
> *Deuteronomy 33:27*

Rather than ask, why does God send grief, it might be more profitable to ask, what are the results of grief? For each person the results will be different, but the desire of God is the same for every person.

While some will collapse in hopeless despair, the purpose of despair is that we collapse in total dependence upon God. Why else would God have promised to be our refuge if He hadn't known we would need a hiding place? Why else would He have promised everlasting arms to uphold us if He hadn't known we would need the security of His love and protection?

When grief drives us in dependence to God, we become stronger persons because we have plunged to the depth of our weakness. We become more feeling because we have felt intense pain in the deepest part of our being. We reach a higher level of maturity because we have developed an intimate relationship with our Refuge.

Grief does bring positive changes when we accept what God has taken and given.

> How should man be just with God?
> If he will contend with him, he cannot answer him one of a
> thousand. He is wise in heart,
> and mighty in strength: who hath hardened himself
> against him, and hath prospered?
> *Job 9:1-3*

Do you feel life is just too painful to go on living? This world is disappointingly full of imperfect, troublesome people and heartrending events.

We would like our days to be cloudless and our nights to be serene. Too often, we humans think the peace and joy of Christian living ought to eliminate times of struggle and sorrow, rather than merely support us during such times. When "my heart doesn't overflow with wondrous delight," I contend with God. "Why is life such a pain?" Somehow it's easy to forget that the flawless, perfect place is not on earth—it's only in heaven.

Lord, help me to be realistic about this sinful world
and fill my heart with the divine art of comfort!

> I AM POOR AND NEEDY; YET THE LORD THINKETH UPON ME.
> *Psalm 40:17*

In high school, I was placed in a class called DP instead of PE because I had had rheumatic fever and couldn't exercise a lot. To be a DP was to be in a class all of its own. There were others in the class with various maladies or other reasons for not being able to participate in PE. But we all had one thing in common—we were abnormal in some way. Because of our common lot, we found it easy to relate to each other.

How often my grief and loss has caused me to feel like a displaced person! I just don't fit in "normal" society any longer. When you become a widow, a widower, a childless couple, or minus a child, or experience any other major loss, you become a different person. You belong in the category of the abnormal.

Sometimes it is easy to classify yourself as abnormal when there are many other "abnormals" around you. But it is another story when you are surrounded by "normal" people. This is the time you need to remember and rejoice with the Psalmist, as in verse 17 of Psalm 40.

Trust God where you cannot trace him. Do not try to penetrate the cloud he brings over you; rather look to the rainbow that is on it. The mystery is God's; the promise is yours.

—Macduff

> **MY TEARS HAVE BEEN MY MEAT DAY AND NIGHT,**
> while they continually say unto me, Where is thy God?
> *Psalm 42:3*

Everyone who has experienced grief has a story to tell. Truly they can answer that question, "Where is thy God?" For they have found Him in the deepest, darkest night of their life.

Though painfully heartrending, it is a story of a shattered life being held together by God's love and grace. It is a story of a dark, lonely valley which is slowly becoming alight with the glow of God's guidance. It is a story of hopeless despair turning into total God-dependence. It is a story of tears, broken dreams, and desperate feelings of being forsaken. It is a story of being carried when rigid with pain, of being understood while in a state of bewilderment, of being sustained when life was meaningless.

God wants to write a story of hope, goodness, and grace in each person's life, but the life of the grief-burdened, broken heart that has found healing tells it best.

Yours is a story worth telling. Share your heartbreak and grief-healing experiences with someone!

> I KNOW THY WORKS, THAT THOU ART NEITHER COLD NOR HOT:
> I would thou wert cold or hot, so then because thou art lukewarm, and neither cold nor hot, I will spue thee
> out of my mouth.
> *Revelation 3:15, 16*

These verses express a passionless spirituality. Too often, Christians feel they must express a passionless grief.

Lord, I weary of trying to maintain a false front which says, "Everything's fine." I'm drained because of that facade of peace and acceptance people expect me to express.

I need to express my intense emotions of anger and frustration because of my loss, my anger at the horrid inward pain, my anger at a seemingly absent God. Help me, Lord, to deal with these emotions in a way that will bring healing, not guilt. My emotions are not lukewarm, they're hot.

My thoughts and emotions are abnormal right now, but I know in my "normal" mind that God is big enough to handle my anger. He is strong enough to cope with my near-insane thoughts. He is wise enough to comprehend my human ignorance and fragility. He understands my heart, even being glad enough for my exasperated expressions that He will not spew me out of His mouth! He will hold me to His heart!

Thank You, Lord, for the strength and love of Your understanding.

> AND NOW, LORD, WHAT WAIT I FOR?
> my hope is in thee.
> *Psalm 39:7*

Hope is activated by hurt and disappointment, pain and sorrow. When life is going great, what is there to hope for? So, hope is a "grief" word and also a "love" word.

When the bottom has fallen out of my world, when my life is chaotic with pain, when my dreams lay in broken pieces at my feet, I ask, And now, Lord, what wait I for? When nothing is left, that's when hope steps in the door.

Hope in what?

   Hope in whom?

      Hope in the Lord!

My soul, wait thou only upon God; for my expectation is from him (Psalm 62:5). When I wait for God, I am expressing faith in Him. When I express faith in Him, He will work!

My expectation is in God because power belongeth unto God (Psalm 62:11). What do I want from Him? What is He going to do for me? How is He going to help me?

Only He knows what is best for me. Can I trust the "what" and "how" to His decision?

My hope is in Thee!

> WHAT IS MAN, THAT THOU SHOULDEST MAGNIFY HIM?
> and that thou shouldest set thine heart upon him?
> *Job 7:17*

Grief challenges you with the discovery of things about yourself that you never really knew before. Grief stretches every inch of your being, making you feel vulnerable. Yet, through pain, you become enlarged with new growth in new areas.

It pushes you out of your comfort zone into extreme discomfort and self-absorption. At first, you can feel and deal only with your own pain. As time moves on, you become sensitive to the pain of others.

It opens a raw and bleeding heart to the spectators, who wonder in amazement as the gradual healing reveals a more gentle and caring you.

Grief makes you feel like a spectacle, and at the same time it makes you feel like a recluse. You discover strengths and weakness you were unaware of before.

God's eye and heart was set on Job for good, but Job felt only the huge sore that was his soul and the burning in his body.

When God measures a man, He puts the tape
around his heart, not around his head.

> HE IS THE ROCK, HIS WORK IS PERFECT:
> for all his ways are judgment:
> a God of truth and without iniquity, just and right is he.
> *Deuteronomy 32:4*

This is a verse I must continually keep in mind as I journey through any great trial. When I focus on a perfect God who does all things right, I will heal more quickly and more correctly. (Yes, it is possible to heal incorrectly!)

Without His love, I would feel no motivation to go on living. Without His spirit, I would be lost in the dark valley of grief. Without His promises, I would lack encouragement and inspiration to pick up and try again. Without Him, there would be no deliverance from the valley and no hope for a brighter tomorrow.

Knowing Christ does not create nor necessarily curb the sorrows and trials of life; but my relationship to Him cushions my heart and connects my mind to a greater power than itself. From Him flows strength to endure, grace to enlighten, and wisdom to emerge a stronger and wiser person.

Oh, Lord, that is my desire!

A man who is willing to be taught is in better condition than a man who thinks he is able to teach.

> OH, THAT I MIGHT HAVE MY REQUEST; . . .
> Even that it would please
> God to destroy me; that he would let loose his hand,
> and cut me off!
> *Job 6:8, 9*

How many times have I thanked God for not giving me my requests, for not answering my prayer of despair? I'm certain Job was later very thankful that God didn't give him the many things he requested in his desperation. Thank You, God, for Your compassionate, empathizing understanding.

Job knew he was in God's hand and didn't consider taking his own life, though death would have been a pleasure in place of the life he was enduring.

Thank You, God, for Your hand that holds my life. After my prayer of desperation and my prayer for deliverance, help me to realize the wiser prayer is to plead for Your grace to endure.

How do I interpret what You bring into my life and into the lives of others? In the lives of others, do I call their trials discipline? Do I consider my valley experience as a trial sent by Satan and allowed by God? Perhaps I should leave the interpretation up to You!

Prayer is not overcoming God's reluctance;
it is laying hold of His highest willingness.

> PROVE ME . . . SAITH THE LORD . . . IF I WILL NOT open you the windows of heaven, and pour you out a blessing.
> *Malachi 3:10*

In Italy, there is a mansion with 365 windows. The builder's intent was that there should be a different window to look out at a different angle on the scenery every day.

When one is hurtled into the valley of grief, there seems to be no windows and positively no different perspective for each new day. All you see, feel, smell, hear, and taste are tears, pain, and sorrow. When you finally do look out a window, what you see reminds you of your missing loved one. What you touch, smell, hear, and taste simply brings back painful memories.

But keep looking, touching, smelling, hearing, and tasting! Slowly a window will open, and you will see others hurting as badly as you are. Another window opens, and you feel the compassion of another hurting friend. Another window opens, and you smell the sweetness of flowers offered to you in love. Another window opens, and you hear the Savior weeping with you and for you. Another window opens, and you taste the goodness of the Lord after the bitterness.

Keep opening new windows of thanksgiving, forgiveness, and healing.

> THE LORD IS MY ROCK, AND MY FORTRESS, AND MY DELIVERER;
> my God, my strength, in whom I will trust;
> my buckler, and the horn of my salvation.
> *Psalm 18:2*

"The Lord is my rock"—

    when my small world around me crumbles.

"My fortress"—

    when troubles and trials are behind and before me.

"My deliverer"—

    when Satan has cast down my heart with fear and hopelessness.

"My strength"—

    when I am weak with the pain and sorrow of grief.

"My buckler"—

    when Satan's darts of discouragement fly thick and fast.

"The horn of my salvation"—

    when my soul is overwhelmed.

"My high tower"—

    when my tears threaten to flood out all joy and peace.

"MY God, in him I will trust"—

    for there is no other god more personal, no other being more trustworthy, no other help more secure, no other heart more compassionate, no other comfort more supportive.

A fortress without God, is like a ship without an anchor.

> REMEMBER THEM THAT ARE IN BONDS,
> as bound with them; and them which suffer adversity, as being
> yourselves also in the body.
> *Hebrews 13:3*

I know of no darker, more lonesome prison than that of grief. Who remembers those in bonds? Who suffers along with those in adversity? How do you remember and suffer along with the hurting?

One of the best and truest definitions of sympathy is, "Sympathy is *your* pain in *my* heart." It is more than feeling sorry for the suffering one. It is more than viewing their distress at a distance.

True Christian sympathy requires that we bring ourselves into personal contact with the hurting. To bear another's burden is to become personally involved in their pain.

The valley experience can make you like a seismograph—the instrument that measures the slightest and greatest tremors and shocks of earthquakes—or it can leave you deaf and dumb to the world around you. The end results depend on your ability to learn from experience and to accept God's plan for your life.

Lord, make me sensitive to the sights and sounds of human need.

> HAVE MERCY UPON ME, O LORD, FOR I AM IN TROUBLE:
> mine eye is consumed with grief, yea, my soul and my belly.
> For my life is spent with grief, and my years with sighing: my
> strength faileth because of mine iniquity,
> and my bones are consumed.
> *Psalm 31:9, 10*

Anyone who has experienced grief knows the "belly" is the very center of physical life and of the emotions. You feel the pain right in your middle!

David, unfortunately, was the cause of much of his own grief. His life was full of bitter reaping as the result of his adultery with Bathsheba and consequent murder of her husband, Uriah. But David knew where to go to find forgiveness—forgiveness that only God's mercy can offer.

In grief, there is often the need for forgiveness, forgiveness of self, of others; sometimes even forgiving the departed one is necessary. Grief cannot heal without forgiveness. Examine your heart to see if forgiveness is needed in any area of your life.

Forgiveness is the key which unlocks the door of resentment and the handcuffs of hatred. It breaks the chains of bitterness and the shackles of selfishness.
—Corrie Ten Boom

Forgiveness is a matter of choice—not feeling.

> CAST THY BURDEN UPON THE LORD,
> and he shall sustain thee."
> *Psalm 55:22*

Have you heard the story about the blue jay who fought and fussed at his reflection in a shiny hubcap? He was his own worst enemy and didn't even know it!

You may be the biggest obstruction to your own grieving process. When you experience all those negative grief feelings—confusion, anger, loneliness, indecision, detachment, exhaustion, etc., where do you go with them?

The grieving person often finds it difficult to follow the direction given in Psalm 55:22. *"Cast thy burden upon the LORD, and he shall sustain thee: he shall never suffer the righteous to be moved."* So many times that "sustaining" needs to be done by a "flesh and blood somebody." Often, the widow in particular, doesn't have that somebody. She may find herself feeling like one widow expressed it:

I feel like I'm playing a losing game of Monopoly. It's pay rent here, pay rent there, go to jail, pay your taxes, and pay the player next to you. Someone else always gets all the breaks. There's not enough of me or money to go around.

Now is the time to remember to give your burden to God,
but you won't cast it there if you don't truly believe
and trust that He is able to deliver.

> I CRIED UNTO GOD WITH MY VOICE,
> even unto God with my voice;
> and he gave ear unto me. In the day of my trouble I sought the Lord: my sore ran in the night, and ceased not: my soul refused to be comforted. I remembered God, and was troubled:
> I complained,
> and my spirit was overwhelmed. Selah.
> *Psalm 77:1-3*

A difficult, but true fact for me to accept after my husband was diagnosed with dementia was this statement: Sometimes God gives us more than we can handle so we will share the load with others. Remember what Jethro told Moses:

> And it came to pass on the morrow, that Moses sat to judge the people: and the people stood by Moses from the morning unto the evening. And when Moses' father in law saw all that he did to the people, he said, What is this thing that thou doest to the people? why sittest thou thyself alone, and all the people stand by thee from morning unto even? And Moses said unto his father in law, Because the people come unto me to enquire of God. . . . And Moses' father in law said unto him, The thing that thou doest is not good. Thou wilt surely wear away, both thou, and this people that is with thee: for this thing is too heavy for thee; thou art not able to perform it thyself alone (Exodus 18:5, 17, 18).

Then Jethro told his son-in-law to get other men to help him. He needed to delegate some of his authority to others.

It is not easy to ask for help, but sometimes it is very necessary.

Let's remember it is good to swallow your pride—it is non-fattening.

> **LIKE AS A FATHER PITIETH HIS CHILDREN,**
> so the LORD pitieth them that fear him. For he knoweth
> our frame; he remembereth that we are dust.
> *Psalm 103:13, 14*

An eagle spied a beautiful, big salmon in the river. With one great swoop, he descended and clutched the big fish in his talons. Then, flapping his huge wings, he started lifting the fish from the water. Several times the eagle tried to rise, but he kept losing altitude. The fish was too big.

The eagle struggled to loosen the fish from his grasp, but it was caught in his talons. Slowly, the grand bird was pulled into the rolling river, becoming prey of his prey.

That story struck a chord in my heart. I was just like that eagle, trying to lift and carry a load much too big for me. Indeed, I had begged God hundreds of times to help me, yet the load was too big. I was being carried downward by my pride, guilt, and unwillingness to accept my humanity. I was my own biggest obstacle in my grieving process and my getting on with the living process.

*I needed help—not only the help of God, but the
help of some human hands and hearts.*

> M<small>Y</small> <small>TIMES ARE IN THY HAND</small>: <small>DELIVER ME</small> . . .
> make thy face to shine
> upon thy servant: save me for thy mercies' sake.
> *Psalm 31:15, 16*

My times are in Thy hand; therefore I will cling to Thee when the waves of grief surge over me. My times are in Thy hand; therefore I must trust that nothing can happen in my life but what God orders. My times are in Thy hand; therefore I will be content and not fret.

It has been said, "The stream cannot rise higher than its source." Nor can my faith and trust rise higher than its source. When my faith and trust is in God, I have the highest source with the greatest possibilities! When my faith is in myself and my abilities, I am at the lowest source with the least possibilities!

Help me, Lord, to keep my times in Your hand. Help me to remember that a lifetime is not made up of weeks, months, and years, but of breaths and heartbeats. Help me to feel the pulse of other needy travelers around me.

Suffering is a gift we can give back to God
to prepare us for our eternal home.

> SO NOW ALSO CHRIST SHALL BE MAGNIFIED IN MY BODY,
> whether it be by life, or by death.
> *Philippians 1:20*

In this verse, we see that Paul had come to grips with death. It mattered not whether he lived or died; what mattered was that Christ be magnified through him.

Grief brings you face to face with your own mortality. Suddenly, you realize you cannot live well until you can die well.

Following is an observation printed in a New Zealand newspaper:

Cancer makes people start thinking about the quality of their lives. Everything they do has a keener edge on it and they get more out of life. In fact, some people never become completely human beings and really start living until they get cancer. We all know we are going to die some time, but cancer makes people face up to it…They are going to go on living with a lot of extra enjoyment, just because they have faced the fear of death. Cancer patients aren't dying. They're living. I have never seen a suicide because of cancer.

Samuel Johnson said, "When a man knows he is to be hanged in the fortnight, it concentrates his mind wonderfully!"

Certainly, all the issues of life and death come into sharper focus when one's future is uncertain.

The key to life is coming to terms with death.

> BE OF GOOD COURAGE, AND HE SHALL STRENGTHEN YOUR HEART, all ye that hope in the LORD.
> *Psalm 31:24*

God offers no promise to shield us from the evil of this fallen world. There is no immunity guaranteed from sickness, pain, sorrow or death. What He does pledge is His never-failing presence for those who have found Him in Christ. Nothing can destroy that. Always He is with us. And, in the long run, that is all we need to know.
—J. I. Packer in Fear No Evil

"Sorrows are our best educators." Make note of what your sorrows teach you!

"A man sees more through a tear than a telescope." While you can view the wonders of the heavens through a telescope, tears reveal the inside of your soul.

"The soul would have no rainbow, had the eye no tears." The rainbow is the promise of God's faithfulness to keep His promise. That's something you cannot live without. His faithfulness to keep His promises is the guiding beacon through the dark valley.

Every divine promise is built upon four pillars!
God's justice or holiness,
which will not suffer him to deceive; His grace or goodness,
which will not
suffer him to forget; His truth, which will not suffer him to change;
and His power, which makes him able to accomplish.
—Salter

> **LET ME DIE THE DEATH OF THE RIGHTEOUS,**
> and let my last end be like his!"
> *Numbers 23:10*

In dealing with grief, one cannot help but think of the last words of many dying people. Some have been words of hope, some of despair, but every word an indication of how that person lived.

An eighteenth-century society figure, Lady Mary Wortly Montague's last words were, "It has all been very interesting."

In 1603, Queen Elizabeth at the age of 70, after cruelly persecuting the Puritans, said, "All my possessions for a moment of time."

The famous man of history books, Thomas Paine, was cruel and unhappy. His wife died because of ill usage. He died in disgrace and poverty. His last words were, "God, why have you forsaken me?"

Frances Havergal, poet and hymn writer, was happy when her doctor said she would soon die. She replied, "Beautiful, too good to be true. Splendid to be so near heaven's gate."

Susanna Wesley, mother of 19 children, made her dying request, "Children, as soon as I am released, sing a psalm of praise to God."

Consider some of the last words of Christ, *Father forgive them, for they know not what they do* (Luke 23:34), and the words of the thief on the cross, *Remember me when thou comest into thy kingdom* (Luke 23:42).

> *Let us be sure we can say, "If I die, I shall be with God;*
> *if I live, He will be with me."*

> IT IS GOOD FOR ME THAT I HAVE BEEN AFFLICTED;
> that I might learn thy statues.
> *Psalm 119:71*

Death has a way of making time stand still—a way of pulling the mind out of time into eternity. Life rushes on about me, but I feel it not. What was important to me yesterday no longer matters today. I stand face to face with eternity—with my own mortality. My time to die is also in the making.

The glitter and glitz of the world about me fails to attract and charm me. The eternal light of heaven has become much more real to me in this dark valley of grief.

Help me, Lord, to remember that You have an infinite awareness and knowledge of all that concerns me. During this grief, and at all times, You love me, You watch over me. And more than that, You have a purpose and plan for me.

Right now, Your purpose and plan is totally obscure to my tear-strained eyes. Yet, help me to hold onto the facts and move on through my feelings to arrive at security in Your love.

We learn much less in life's school of affluence
than we learn in life's school of adversity.

> MY SOUL IS WEARY OF MY LIFE;
> I will leave my complaint upon myself;
> I will speak in the bitterness of my soul.
> *Job 10:1*

In this verse, Job implies that, up to this point, he has put some restraint on his expressions; now he is ready to dump all his complaints. Job is weary of the pain in his life and is ready to speak his bitterest feelings.

This chapter is the wail of a crushed heart. As he cries out to God, Job expresses his feelings honestly. He feels condemned and despised by God. He is in perplexity, confusion, rejection, and hopeless despair.

Fortunately, Job doesn't feel compelled to follow the path of idealism set by his non-grieving friends. But his realistic expressions of grief shock his friends and cause them to accuse and mock Job even more. Idealism belongs only to non-grieving individuals!

Grief is so painful and wearisome, so exhausting and horrible, so shocking and stressful that it must be given words. Be realistic with God. He can handle raw honesty better than your friends can. God isn't threatened by your honesty. Realism facilitates healing.

God is the master mender of broken lives.

> THE TROUBLES OF MY HEART ARE ENLARGED:
> O bring thou me out of my distresses. Look upon mine affliction
> and my pain; and forgive all my sins.
> *Psalm 25:17, 18*

How would you feel if your loss was compared with the loss of a friendship? How would you feel about the person who tells you they often think of you, call you, or send you a fax—but only in their mind? Yet, you hadn't the faintest idea they gave you a thought.

As we travel through the valley of grief, we will find many obvious and unexpected ways to exercise a forgiving spirit. Those who have never experienced grief's deepest tragedy truly need compassion and forgiveness for their ignorance and inexperience—just as we so often needed forgiveness before our valley experience.

Comparing the hurt of a broken friendship with the grief of the loss of a loved one is like comparing a broken leg with the amputation of a leg. Broken friendships and broken legs usually can be repaired—though maybe not as good as new. But the loss of a loved one and the loss of a leg cannot be brought back to normal condition. The loss is painful and permanent. There is no comparison!

The greatest misfortune of all is, not to be able to bear misfortune.

—Bias

> SUFFERING IS ONE OF THE FORMS AND MANIFESTATIONS OF Divine goodness in the education of human beings.

Do I hear you exclaim, "Divine goodness?" What can be good about suffering? What can be good about grief? What can be good about pain?

If you do not allow your valley experience to make you a more compassionate person; if you do not allow God to make you more sensitive to hurting people about you; if you do not ask God to give you more understanding and the ability to weep with those who weep; if, in your heart, you do not find God greater, stronger, and more faithful than you have ever known Him to be; if heaven isn't nearer and life isn't dearer—then go ahead and exclaim, "Divine goodness?!"

It is only after we accept the fact of God's divine goodness that we can accept the pain of grief. Only then can we let suffering educate us the way God planned!

Is it not the knowledge and hope of a better life that gently leads the mourner through the dark valley? One can only imagine the blackness experienced by those without such a knowledge and hope.

If your life is not anchored in the character and knowledge of God, then grief and loss will be a very dark, hopeless experience indeed!

Afflictions are a token of God's desire to exalt man.

> TURN THEE UNTO ME, AND HAVE MERCY UPON ME;
> for I am desolate and afflicted. The troubles of my
> heart are enlarged: O bring thou me out of my distress.
> *Psalm 25:16, 17*

You cannot intellectually think yourself through, around, over, or under the grief process.

You should not bury your grief.

You should not resist your grief.

Grief is not abnormal, optional, or a sign of spiritual weakness.

Grief is normal, natural, and necessary.

Grief is exhausting, excruciating, and explosive.

Grief can be instructional and beneficial or devastating and detrimental, depending on how you deal with it.

You must feel your way through grief.

You need to cry your way through grief.

You need to talk your way through grief. You must accept and embrace your grief. You must pray and pray your way through grief.

Your pain and grief are uniquely yours. Your valley is your valley to walk with the Man of Sorrows alone.

*We tend to avoid our greatest blessings
because they come wrapped in difficulties.*

> THE PRESENCE OF THAT ABSENCE IS EVERYWHERE.
> —*Edna St. Vincent Millay*

The scent of a candle, the color of a tulip, the sight of a butterfly, the song of a bird, the sound of laughter, the smell of the house, the buzz of a bee, the taste of the pie, the sight of a rainbow, the clap of thunder, the clouds in the sky, the moonlit night all remind me of that absence.

The strong oak tree, the sound of a tractor, the roar of a motor, the sight of a fishing pole, the sound of the alarm, the crowing of the rooster, the slippers by the chair all remind me of that absence.

That absence is everywhere. It screams at me, it calls to me, it cries out to me, it shouts at me, and it whispers to me.

What shall I do, Lord, with the presence of that absence?

How shall I fill that empty spot in my heart?

I will be to him [me] a father [a mother, a sister, a brother, a son, a daughter], and he shall be to me a Son (Hebrews 1:5).

Thank You, Lord, for my family connections with You!

> MY HEART IS SORE PAINED WITHIN ME:
> and the terrors of death are fallen upon me. Fearfulness and trembling are come upon me, and horror hath overwhelmed me.
> *Psalm 55:4, 5*

Grief affects you emotionally. You may feel sad, miserable, depressed, empty, and lonely. You may feel guilty, angry, confused, and forsaken. You may have such a jumble of feelings that it is difficult to identify any one of them. The emotional effects of grief are the most obvious and will produce the most comments from your friends.

Grief affects you spiritually. Some people feel closer to God, and some feel farther away from Him. Some sense God's presence right beside them, and others feel God has turned His back on them and is punishing them.

Suddenly, you are more aware of death and eternity than of life and material possessions. As a Christian works through his grief, he usually grows in faith and develops a closer relationship with the Shepherd who walks the valley with him.

Pain falls drop by drop upon the heart
until, in our despair, wisdom comes.

> AND I SAID, OH THAT I HAD WINGS LIKE A DOVE!
> for then would I fly away, and be at rest.
> *Psalm 55:6*

Grief affects you socially. If you lost a mate, you will relate to others in a very different manner. You no longer fit with the couples, nor do you fit with the singles. You are a class by yourself. You are alone and lonely. You may be overwhelmed with added duties and new responsibilities. You feel like a mole digging in dark tunnels, not knowing where you are going or where you will end up.

Grief affects you physically. You may have trouble sleeping. You may fear death yourself. You may experience panic or "heart spells." You may eat too much or not enough. You may have pain all over. You may suffer fatigue and lack of motivation.

Grief affects you mentally. You may feel like you are losing your mind. You can't remember well and find it difficult to make decisions. Your mind may feel dazed or foggy. Your mind may be confused, and you find it difficult to concentrate.

After reading these effects of grief, you can say: Now I know I am a normal mourner.

He will no load of grief impose, Beyond the strength that He bestows.

> SO THEY SAT DOWN WITH HIM UPON THE GROUND
> seven days and seven nights, and none spake a word unto him:
> for they saw that his grief was very great.
> *Job 2:13*

Job's visitors have been rightly called "miserable comforters." But, in all fairness, we should recognize their good points also.

They proved themselves to be friends by going to see Job. We would like to think that trouble collects friends, but the opposite is truer. Friends disappear in the hour of trouble.

Job was in deep trouble, and only three genuine friends chose to expose themselves to his pain. Someone has said, "It is well to go to the house of mourning. But few are they who know how to conduct themselves when there."

Job's friends, unable to imagine his great distress and grief, were overwhelmed when they saw him. Their hearts were filled with sympathy, but their tongues were speechless. For seven days and nights they sat with Job—without words.

At this point, we learn that no words
are far better than painful words.

> HE THAT KEEPETH HIS MOUTH KEEPETH HIS LIFE.
> *Proverbs 13:3*

The friends of Job came with good intentions, but failed miserably when they tried to figure out God's reasons and Job's sins.

When I try to figure out the "whys" and "wherefores" of God's dealing in my own life or in the lives of others, I am setting myself up as an all-knowing judge. I am assuming the place of God.

How much better it would have been if Job's friends would have accepted the sovereignty of God in faith, without their intellectual reasoning and questioning.

A squeeze of the hand, a hug, a mingling of tears, a look of love means more than a hundred words.

Though Job's friends said many right words, they were inappropriate and caused much pain and injury at the time.

Oh, God, preserve me from such comforters and from being such a comforter. May I encourage, not discourage. May I help, not hinder. May I help decrease the pain rather than increase it. Help me to keep my mouth, so that I may keep my life and the life of my friend.

> *Sorrow can never conquer, nor grief despair,*
> *When there are others who care.*

> HE BROUGHT ME UP ALSO OUT OF AN HORRIBLE PIT, out of the miry clay, and set my foot upon a rock, and established my goings. And he hath put a new song in my mouth, even praise unto our God: many shall see it, and fear, and shall trust in the LORD.
> *Psalm 40:2, 3*

Lord, is it possible that after my grief experience I will sing a new song? Not a song of relief from grief or escape from grief, but a song of deliverance through grief. Deliverance made possible through God's love, goodness, and care!

Yes, Lord, You have promised to bring me up from this pit of tumult and misery. You have promised to deliver me from the miry clay of doubt, pain, and fear. You have promised to make my steps firm—that is security and guidance from You. And You are going to put that new song on my lips.

Because of Your work in my times of tribulation, many will be challenged to trust in You.

Thank You, Lord, for the deliverance that is coming and more that is yet to come!

Walking by faith means putting one foot in front of the other—with both eyes on God.

> FOR I AM READY TO HALT, AND MY SORROW IS
> continually before me.
> *Psalm 38:17*

Do you feel like David, weak and helpless and about to stumble and fall at any moment?

Sometimes our loss and great pain is the result of our own sin, as in David's case. But many times grief strikes us as a bolt of lightning—suddenly and uninvited. We stagger and almost fall, trying to figure out why God has so afflicted us.

The "why" of my suffering is continually before me. Oh, God, help me not to fall into the pit of doubt and bitterness. Help me to accept what has come, realizing all that comes my way has first passed by You.

For some reason You have put me in this hard place. Lord, make me to know my end and the number of my days. Help me to live them out wisely.

I am frail, but You are strong.

  I am bewildered, but You are discerning.

    I am foolish, but You are wise.

      I am fearful, but You are courageous.

        I am helpless, but You are my help!

Only the heart that suffers can know its own bitterness.

> HE HEALETH THE BROKEN IN HEART,
> and bindeth up their wounds.
> *Psalm 147:3*

Now I fully understand the meaning of brokenhearted. My heart is in pieces—shattered, but not destroyed; crushed, but not ruined. One piece is missing, yet this verse says God will heal and bind together the broken pieces.

I can't put my life back together, I can't stick my heart together again, I can't bind up my emotions, I can't accomplish the healing God has promised. How is this miracle going to get completed?

Help me to remember:

» *The Lord is nigh unto them that are of a broken heart* (Psalm 34:18).

» *Oh, how great is thy goodness* (Psalm 31:19).

» *Jesus was* [is] *moved with compassion* (Matthew 14:14).

» *I will answer thee, and shew thee great and mighty things, which thou knowest not* (Jeremiah 33:3).

» *I am the Lord . . . is there any thing too hard for me?* (Jeremiah 32:27)

» *I will strengthen thee, yea, I will help thee* (Isaiah 41:10).

Somehow, in God's time, with my surrender, the adhesive of God's nearness, goodness, compassion, and mighty strength will fuse my heart back together.

> *God without man is still God;*
> *Man without God is nothing.*

> IT IS OF THE LORD'S MERCIES THAT WE ARE NOT CONSUMED,
> because his compassions fail not. They are new
> every morning: great is thy faithfulness!
> *Lamentations 3:22, 23*

If you read through the first three chapters of Lamentations, you will find these words—some of them listed several times. Can you relate to these words of the "weeping prophet"?

Solitary, weepeth, tears, none to comfort, sighs, bitterness, affliction, miseries, despised, sorrow, desolate, faint, trodden under foot, crushed, eye runneth down with water, distressed, troubled, heart is turned within me, grievous, heart is faint, swallowed up, thrown down, cut off, destroyed, cast off, abhorred, lament, languished, broken, keep silence, dust and sackcloth, hang down their heads, bowels are troubled, liver is poured out, swoon, who can heal thee, burdens, banishment, they hiss and wag their head, they gnash the teeth, heart cried, cry out in the night, pour out thine heart, terrors, consumed, against me, dark places, hedge me about, made my chain heavy, I cry and shout, made my paths crooked, derision, wormwood, my hope is perished, and humbled me.

Israel's terrible distress was the result of their own sin. Your distress is likely the result of living in a sin-cursed world. Yet, in the midst of this great distress, God promises His compassion, mercy, and grace.

He is faithful!

> SURELY EVERY MAN WALKETH IN A VAIN SHOW: surely they are disquieted in vain: he heapeth up riches, and knoweth not who shall gather them. And now, LORD, what wait I for? my hope is in thee.
> *Psalm 39:6, 7*

An archbishop was met by a friend who asked, "You have been to hear a sermon?"

The bishop replied, "I have met a sermon—for I met a corpse, and right and profitable are the funeral rites performed, when the living lay it to heart."

Oh, yes, a corpse is a sermon. His silence bids me ponder the path of my life and the value of my words. He illustrates the vanity of life. All that he worked hard for is left behind. His dead body also reminds me of the frailty of life—here today and gone tomorrow.

The funeral abruptly brings me face to face with the fact that life is full of sorrow and disappointment when there's no hope in God.

Shakespeare expresses much the same sentiment as David in verse six. "Out, out, brief candle, life's but a walking shadow, a poor player that struts and frets his hour upon the stage, and then is heard no more."

Such is not our life, or our end!

Thank God for Christ our hope!

> OUT OF THE PRESSES OF PAIN
> Cometh the soul's best wine;
> And the eyes that have shed no rain
> Can shed but little shine.

Why is life for some so full of disappointments, sorrows, and suffering, while others seem to travel the path of ease and fulfillment?

For those of us who seem to have more than our share of pain and grief, it is better if we ignore the "why" question and focus on the "who" question.

My Father is the husbandman—not Satan, but God the Father. Even Jesus submitted to the suffering God allowed in His life.

The spirit of resentment and self-pity cannot overwhelm us when we remember that God is the husbandman, pruning away what is detrimental in our lives and making growth and fruitfulness possible.

We may cry out during that painful pruning, but it is those tears of submission that will add the shine to our eyes.

The growth in our lives will help others desire
to become what they see us becoming.

> AND BESIDE THIS,
> [because of God's exceeding great and
> precious promises] giving all diligence, add to your faith...
> *2 Peter 1:5*

In the above passage, Peter assumes and knows we have faith because of what God has done for us. And, because of our faith, he prods us to add other Christian virtues.

Have you ever thought about the valley experience as being a time to add? How can we add when we have lost?

It will surely take diligence to add to our faith in the valley of grief, but it is possible if we focus on God's all-sufficient grace rather than our all-consuming grief.

The grief experience is a real testing time for faith. But it certainly is not a time to lose faith, because if we lose faith, all is lost, and there can be no adding experiences.

While journeying through the valley, can we believe:

God is good?
  God is just?
    God is right?
      God is love?
        God cares?

As we work through each question, seeking and trusting His promises, our faith will grow.

Sensing His goodness and compassion will become an adding experience in the valley.

> **ADD TO YOUR FAITH VIRTUE.**
> *2 Peter 1:5*

Virtue has been defined as "active courage."

It takes much courage to move forward in the Christian life! We are told, *Quit ye, like men be strong* [be courageous, stable, manly] (1 Corinthians 16:13).

And oh, how much more courage it takes to move forward through the valley of grief! Grief is a time when we lack motivation, energy, inspiration, and courage!

But the command is to add to your faith active courage. Active courage doesn't simply mean to stand strong and brave at the graveside; it means to take one painful step after another over the unfamiliar, uneven, and unstable dark terrain of the valley.

Only by faith in my compassionate Shepherd am I able to move ahead with active courage.

When I am preoccupied with the past and fearful of the future, I must flee to the Rock that is higher, stronger, and more secure than I.

You cannot know where you're going if
you're always looking back.

> ADD . . . TO VIRTUE KNOWLEDGE.
> *2 Peter 1:5*

What kind of knowledge has my grief experience added to my Christian character?

Grief has clearly defined "heartbreak!" It has taught me the meaning of tragedy, aloneness, and abnormality.

Grief has shown me the value of compassion and empathy. It has exposed those who only know and live the "normal" life.

Grief has declared the necessity of faith, courage, and forgiveness. It has confirmed the sovereignty of God and the limitations of man.

My grief experience has enlightened my mind and tenderized my heart toward God and His grace, the trials of fellow pilgrims, and the judgment of others.

Grief gives a new understanding of God's sufficiency and my personal insufficiency, His strength and my weakness, His care and my dependence.

The knowledge gained by means of grief has been painfully pulverizing, yet bountifully beneficial.

> ADD . . . TO KNOWLEDGE TEMPERANCE.
> *2 Peter 1:6*

Our knowledge and behavior must be in subjection to temperance in every area of the Christian life or we become hypocritical. Temperance—the governing grace of God!

Our appetites, desires, affections, and tempers without self-restraint can be self-destructive. Often, those overcome by the anguish of a loss throw self-discipline to the wind. They become drug, alcohol, sex, or work addicts trying to ease the pain in their lives.

An area of deep struggle for the grieving Christian may be in the matter of temperate thinking. Do I think it is God's duty to tell me why He took my loved one? Do I think God isn't carrying His share of the huge load He has placed upon my shoulders? Do I ask if my pain, chastisement, discipline, or testing by Satan is allowed by God? Do I question if I will ever learn what God is trying to teach me?

Certainly these questions have gone through my mind, but can I be temperate with my demands on God?

Help me, O Lord, to rest in Your governing grace, remembering that I am but dust, and Your plans for me are good and not evil.

> ADD . . . TO TEMPERANCE PATIENCE.
> *2 Peter 1:6*

Patience is the conscious submission of our human will to the holy will of God. Oh, my grieving companion, isn't that a mouthful? Perhaps it is better to say, a heart full of commitment.

It may be easy to say, "My loved one is in a better place. He/she wouldn't want to come back. This must have been God's will, or He wouldn't have allowed it to happen." Yes, there are a lot of trite things the mouth may say, but honestly what does the heart feel?

How patiently can I endure this dark valley God has brought me into? Can I patiently endure this agonizing process of accepting reality? Can I patiently endure this slow progression of healing? Can I patiently endure my failure to meet people's expectations? Can I patiently endure my sorrow, my loneliness, my confusion, and my adjustment to a "new normal"? Can I patiently endure being ignored, misunderstood, or wrongly judged?

In spite of every surrounding factor concerning my loss, can I submit my will to God's holy will, believing He will work out everything for my eternal good? That is adding patience!

Patience is not passive; on the contrary it is active;
it is concentrated strength.

> ADD . . . TO PATIENCE GODLINESS.
> *2 Peter 1:6*

When God's will becomes my will—that is godliness. Godliness is the spirit of reverence and fear. It is when the thought of God controls the whole of life.

Our sinful, human nature puts godliness on trial throughout our Christian life, but one of the most severe trials is when we face grief.

Perhaps your child was killed by a drunk driver, or was drowned while trying to save another. Perhaps your loved one died of a disease transmitted through a blood transfusion. There are thousands of unfair, unnecessary, unthinkable, and even ungodly ways of losing one we love. Yet, all ways point back to the fact that we live in a sin-cursed world, painfully groaning until the return of Jesus.

Our godly qualities can be tried beyond our limit! At such times, is God's grace sufficient? Are His mercies faithful? Is His understanding unfathomable? Is His compassion immeasurable?

Can I place my trembling hand in His and cry, *Not my will but thine?* That is godliness of the highest hue.

> *Two men please God—one who serves Him with all his heart because he knows Him; and one who seeks Him with all his heart because he knows Him not.*
>
> —Panin

> ADD . . . TO GODLINESS BROTHERLY KINDNESS.
> 2 Peter 1:7

The natural outflow of our love for God is the outflow of kindness to our brethren. Again, the time of grief is often a time of testing my feelings toward my Christian brothers and sisters.

With some, a love and closeness can develop which never existed before the grief experience. Often, this closeness is with others who have also suffered greatly or lost someone dear. With others, an aloofness or strangeness may develop, which had not before existed.

The bereaved may find his brotherly kindness challenged in many new ways.

I have learned that another's estrangement from me is often due to my change in perspective, or how I perceive that person is looking at me. My alienation from others may be because they feel uncomfortable with me and my sorrow.

Some care, but don't know how to show it,

While others don't care, and they let you know it.

Yet it is my godly duty to show brotherly kindness to all. What an opportunity to grow!

God never puts you in a place
too small to grow!

> TO APPOINT UNTO THEM THAT MOURN IN ZION,
> to give unto them beauty for ashes, the oil of joy for mourning,
> the garment of praise for the spirit of heaviness; that they might
> be called trees of righteousness, the planting of the LORD, that
> he might be glorified.
> *Isaiah 61:3*

It may take a long while before mourning is changed into beauty, joy, and praise. But we know this is God's desire for each one who mourns. Because of Christ's resurrection, we "sorrow not as they that have no hope."

Most often, a loss brings negative thinking which works in a downward spiral. Try this circuit-breaker to turn your thinking into an upward climb. Accent the italicized word each time you repeat the phrase.

> *God* is my only source---
> God *is* my only source---
> God is *my* only source---
> God is my *only* source---
> God is my only *source*---

It is God's work and resource alone that will turn your mourning into beauty, joy, and praise. Then you can become a tree of strength, a source of refuge, and nourishment for others.

> *When we are at the end of our resources,*
> *we are just at the beginning of God's.*

> I WILL NOT DOUBT, THOUGH SORROWS FALL LIKE RAIN,
> And troubles swarm like bees about a hive.
> I will believe the heights for which I strive
> Are only reached by anguish and by pain;
> And though I groan
> and writhe beneath my crosses,
> I yet shall see through my severest losses
> The greater gain.
> —*Author Unknown*

"In times of trouble, some people grow wings, others buy crutches."

Which person am I? Have I eased my pain with drugs, with work, or by escape? Or have I kept myself busy hobbling over graves where I have buried my feelings, hidden my tears, and covered up my heartbreak?

Or, have I gone to the Rock of Ages? There I will find Him who told Israel, *I bare you on eagles' wings, and brought you unto myself* (Exodus 19:4).

Help me, Lord, to remember that my trials and heartaches are no accident or happenstance. No suffering is purposeless. You have my eternal profit in view.

Oh, God, help me not to waste my sorrows!

> BUT NOW, O LORD, THOU ART OUR FATHER;
> we are the clay, and thou our potter; and we all are the work of thy hand.
> *Isaiah 64:8*

What does it take to be clay in the Potter's hand? It takes total surrender and trust that God knows best. There can be no tiny wires of self-will, no stones of stubbornness, and no grits of self-greatness.

Creath Davis once said, "God doesn't grow great people with easy circumstances—but most of us would settle for a little less greatness and a little more tranquility."

Do I want peace or perfection? Peace only follows the perfecting work of Christ in our lives. How do I become the beautiful vessel perfectly fit for the Master's use?

Oh, yes, I desire to reflect my Creator's image, but who desires to go through the refining fire?

If God is truly my Father, then He will shape me according to His will. He makes or mars as He pleases.

Can I say with submission, "My clay is in your hands?"

> *As the raindrops yield to the stream,*
> *As yoked oxen work as a team,*
> *As the trees bend with each wind gust,*
> *My weak clay to God I entrust.*
> —Faythelma Bechtel

> IF THERE IS A WIDE GULF BETWEEN YOUR FAITH AND FEELINGS right now because of the hurt and pain you are feeling, that's not hypocrisy—that's honesty.
> —*Joey O'Connor*

People told me about "firsts," but I little comprehended what they were saying . . . until the "firsts" began coming.

The first Thanksgiving was only a month after our daughter's death. I was anxious to get it over with. I planned to share the ritual I had shared in grief class.

As I gave each adult member of our family a red rose and shared a memory of our daughter, we were all in tears. It was a time of gathering closer as a family. It was a small element in beginning the healing process. Now, looking back, I realize how numb I was in body and mind at that time. God's merciful anesthetic was still working in my shocked being. Oh, how good it was not to have any idea of what I yet had to work through!

Faith, feeling, and fact grew farther apart as the days passed. There was a lot of reality to work through by acceptance and submission.

Life's trials are part of God's
educational system to develop faith.

> CAST THY BURDEN UPON THE LORD,
> and he shall sustain thee:
> he shall never suffer the righteous to be moved.
> *Psalm 55:22*

On the first anniversary of Mother's Day, my foundations were shaken! I went to church thinking I could handle the day, as long as I didn't attend the Mother's Day carry-in—a time when the fathers prepared the meal and cleaned up the mess.

My husband never did much in the kitchen, and now he did nothing. He had dementia! I grew to dread Mother's Day.

Well, the sermon was on the perfect mother. The mother who had all the patience and wisdom necessary for raising a family. The mother who was never up and down emotionally. The mother who was always stable and calm. The mother who was there for her husband and children.

I hurried my husband out to the car and cried all the way home. It was bad enough to be without my eldest daughter, but it was more than I could take to realize what a failure I was as a mother.

Right then, I was proving myself very unstable emotionally. A multitude of appalling feelings were charging through my body and brain. Suddenly the doorbell rang, and someone opened the door. I didn't bother getting up from the bed where I lay. I was at the end of my emotional rope.

God alone can use and restore broken things.

> SEND THINE HAND FROM ABOVE; RID ME,
> and deliver me out of great waters.
> *Psalm 144:7*

I was alone. I was useless. I was a failure. Would this pain never heal! Yes, I was nearly drowned in tears of self-pity and pain.

I faintly heard the commotion of our dining room table opening, the table being set, and chairs gathered around. How could I face my family? I knew my son, his wife, and six grandchildren had arrived. They had sacrificed their fellowship at church to spend the day with an unworthy mother and grandmother.

Finally, I forced myself out of the shell I had pulled myself into. I opened my grief-stricken eyes to behold a beautifully set table with a lovely food tree of fruit and sculptured vegetables and an exquisite display of cut cheeses and meats and bowls of crackers and bread. The family had outdone themselves using their new food sculpturing tools, and now all their work and attractive display was being wasted on me—a failure as a mother.

I burst into tears. "Please take your beautiful food to church, I don't deserve all this," I begged the children, yet I was so glad they had thought of me.

"No one deserves it more," my son announced. "Now come, sit down and enjoy it with us!"

What could I say? It's amazing how God rescued me from drowning in my ocean of self-pity, failure, and pain.

Thank You, Lord for pulling me up from despair.
Thank You for showing that You truly do care.

> BUT AS HE WHICH HATH CALLED YOU IS HOLY,
> so be ye holy in all manner of conversation; because it is
> written, Be ye holy; for I am holy.
> *1 Peter 1:15, 16*

To what has God called me? Am I destined to happiness? It certainly doesn't feel like it. Am I destined to fame? I haven't found that either! Am I destined to good health? Not with all the problems I've got!

God has called me to holiness. All His children are ordained to holiness. God's methods for producing this holiness differ from person to person. But one thing we can be assured of is that losses have a purifying effect on the loser who is surrendered to God's will.

There are many different kinds of losses; but is there any with more potential of driving you to or away from the heart of God than the loss of a loved one?

Holiness is absolute purity in the thoughts of my mind, the words of my lips, and the actions of my being. That's a supernatural order for anyone, and doubly so when one is in the throes of grief. It is a work which only God can accomplish, in His time and in His way, but He can accomplish it.

Oh, Lord, make me workable clay in the Potter's hand. Make me a winner through holiness.

Christ is the way to holiness.

> AND JACOB CALLED THE NAME OF THE PLACE PENIEL:
> for I have seen God face to face, and my life is preserved.
> *Genesis 32:30*

Read Genesis 32:24-32. When Jacob met God alone, he became a new man with a new name. Jacob arrived at Peniel a conniver, but he left there *a Prince* with God.

Jacob didn't ask for a wrestling match. He didn't ask to suffer. He didn't ask for a limp that placed him in the "abnormal" category. He didn't ask for a new character nor a new name. He only asked for a blessing.

Did Jacob receive a blessing or blight from God? Was he polished or punished by God? Was he rejuvenated or ruined by God?

We never deliberately ask for pain or grief—it just comes.

We who suffer the trauma of grief may limp for the rest of our lives, but we have a choice. We can accept it as a blessing and not a blight. We can receive it as polishing and not punishment. We can allow it to rejuvenate us rather than ruin us.

Always remember, wounded doesn't have to mean ruined.

Jacob's wound always reminded him Who was in control of his life.

> O LORD, THOU HAST SEARCHED ME, AND KNOWN ME.
> Thou knowest my downsitting and mine uprising,
> thou understandest my thought afar off.
> *Psalm 139:1, 2*

Behind my life the Weaver stands,
and works His wondrous will.
I leave it in His all-wise hands,
and trust His perfect skill;
Should mystery enshroud His plan,
and my short sight be dim,
I will not try the whole to scan,
but leave each thread with Him.

Nor till the loom is silent,
and the shuttles cease to fly,
Shall God unfold the pattern, and explain the reason why
The dark threads were as needful in the Master's skillful hand,
As the threads of gold and silver in
the pattern which He planned.

—Author Unknown

God takes us as we are, tangled strands and all, and spends a lifetime shaping and reconstructing us into His image. Since He knows all the inward workings of my heart and mind, only He knows when to add the dark threads and the gold.

Dear Lord, take up the tangled strands
where we have wrought in vain,
That by the skill of thy dear hands,
some beauty many remain.
—Mrs. F. G. Burroughs

> FOR WE HAVE NOT AN HIGH PRIEST
> which cannot be touched with the feeling of our infirmities; but
> was in all points tempted like as we are, yet without sin.
> *Hebrews 4:15*

Are you struggling to control your feelings? Are you trying to hide your feelings? Are you trying to ignore your feelings?

Try following the advice of James E. Miller who said, "The best way to handle your feelings is not to 'handle' them, but to feel them."

Give your feelings names. Accept your feelings, and accept your humanity. Share your feelings with your all-understanding Creator. He felt forsaken by His God. He was *a man of sorrows and acquainted with grief.*

Share your feelings with your family; it will help you to heal together. Silence will drive you apart. Share with an empathizing friend who may even benefit from your sharing.

Often feelings are best expressed in writing. Write letters to the loved one you can no longer talk to. Put the letters away and read them many months later. You will be surprised at the healing that has taken place.

Write a journal of your journey through the dark valley.

> IN THE FEAR OF THE LORD IS STRONG CONFIDENCE:
> and his children shall have a place of refuge.
> *Proverbs 14:26*

Death is unsettling and distressing. I am facing fears and insecurities that I have never faced before.

Fear fogs the brain and is an energy drain.

It triggers anxiety and inflicts pain.

It empties me until no resources remain.

Rid me, Lord, of this negative fear, and fill me with that positive fear of You that brings confidence and strength.

The fear of the Lord casts out all fear of man, all despairing anticipations of possible evil or hurt, and makes the one who trusts God confident and bold.

Help me to remember that the fear of the Lord affects in a positive way my faith, my wisdom, my purity, my obedience, my reverence, my worship, my security, my courage, my awe of God, and my health and life.

There is no doubt that the fear of God influences the disposition and attitude of the one who is confidently trusting in the guidance and control of a living and loving God. God truly becomes my refuge.

The encouraging words, "do not fear," are found 366 times in the Bible—one for each day, and even an extra for leap year.

> MEASURE THY LIFE BY LOSS AND NOT BY GAIN
> Not by laughter and smiles, but by tears and pain,
> For love's strength standeth in love's overflow;
> And he who suffers most has most to bestow.

Can you imagine:
- » The sorrow of Joseph being hated by his brothers?
- » The anguish of Joseph when he was tossed into the pit?
- » The devastation of Joseph when he was sold into slavery?
- » The grief of Joseph when he arrived in a strange land?
- » The alarm of Joseph when pursued by Potiphar's wife?
- » The exasperation of Joseph when he landed in prison for something he hadn't done?
- » The agitation of Joseph when he was forgotten by the butler?
- » The astonishment of Joseph when he was called from prison to the king's palace?

Joseph's life had been nothing but a series of losses, yet the Bible says, *but the Lord was with Joseph* (Genesis 39:21). God had a plan for Joseph's life, but for many years it didn't look or feel like a very good plan.

God turned Joseph's shattered life into a work of art. He became the savior of a nation and of his own family. Surely Joseph must have been committed to God's will, or it couldn't have turned out so marvelously.

Help me, God, to be committed to Your will in whatever comes!

> O LORD, HOW LONG SHALL I CRY,
> and thou wilt not hear! even cry out unto thee of violence, and
> thou wilt not save!
> *Habakkuk 1:2*

Grief has a way of shattering your ideas about God. As you face the pain, the aloneness, the misunderstandings and aloofness of friends, the questions with no answers, the seeming absence of a compassionate God, as you face all that grief and loss on your doorstep, your previous opinions about God then shatter.

You may feel as C.S. Lewis expressed it in *A Grief Observed*. "Not that I am in much danger of ceasing to believe in God. The real danger is of coming to believe such dreadful things about him."

Oh, yes, I have to admit to feeling that God had forsaken me. God did not really care. God did not understand my pain. God did not send anyone to help when I was at the bottom. God just allowed people to multiply my grief. God's grace wasn't sufficient.

What was wrong with me? I simply wasn't finding God to be what all the pat, trite answers said He was.

Then came the time when I realized my trust in God was not dependent on my understanding of God. I began finding God to be more than any trivial answers made Him out to be.

God became my personal light through the dark valley,
my personal companion, and my personal strength.

> ALTHOUGH THE FIG TREE SHALL NOT BLOSSOM, neither shall fruit be in the vines; the labour of the olive shall fail, and the fields shall yield no meat; the flock shall be cut off from the fold, and there shall be no herd in the stalls: Yet I will rejoice in the LORD, I will joy in the God of my salvation. The LORD God is my strength, and he will make my feet like hinds' feet, and he will make me to walk upon mine high places.
> *Habakkuk 3:17-19*

What the writer seems to be saying is, "Although nothing normal (as we humans think of normal) happens in my life, yet I will trust the Lord."

Indeed, that is a mouthful when my faith in God and in myself has been shattered. I once thought I was a Christian with a strong faith, but my experience with grief shattered my self-image. I found myself a weakling physically, emotionally, and spiritually.

Once my self image was in pieces, God could begin reshaping me into His image. Only when I am weak in myself can I commence becoming strong in the Lord.

When the Lord is my strength, He makes my feet like hinds' feet, able to traverse the steepest, roughest, darkest mountains of life. Only then can I say, "No matter what, I will trust!" Help me, Lord!

Troublous times are leaning and trusting times.

> I HAVE HEARD OF THEE BY THE HEARING OF THE EAR:
> but now mine eye seeth thee. Wherefore I abhor myself,
> and repent in dust and ashes.
> *Job 42:5, 6*

Job's ideas about God and his own personal righteousness were shattered during his agonizing grief experience. Job's mind was flooded with questions about his integrity and God's integrity.

Why doesn't God simply destroy me and get me out of my misery? Why won't God pardon my iniquity and let me sleep in the dust? Why does God hide Himself from me? God has delivered me to the ungodly. Oh, that the Almighty would answer me!

Then God came with His questions!

» Questions Job had no answers for.

» Questions that convicted Job of his ignorance and insignificance.

» Questions that revealed God's unquestionable wisdom and greatness.

» Questions that answered Job's questions.

I abhor myself . . . these are the words of a man who has experienced the depths of pain in his life and then begins to understand the heights of holiness and compassion of his God.

When God shows us Himself and ourselves, we must humbly declare along with Job what he said in Job 42:5, 6. Then our shattered images will come together into a beautiful portrait of love and hope.

Christ Only

When across the heart deep waves of sorrow
Break, as on a dry and barren shore;
When hope glistens with no bright tomorrow,
And the storm seems sweeping evermore;

When the cup of every earthly gladness
Bears no taste of the life-giving stream;
And high hopes, as though to mock our sadness,
Fade and die as in some fitful dream,

Who shall hush the weary spirit's chiding?
Who the aching void within shall fill?
Who shall whisper of a peace abiding,
And each surging billow calmly still?

Only He whose wounded heart was broken
With the bitter cross and thorny crown;
Whose dear love glad words of joy had spoken,
Who His life for us laid meekly down.

Blessed Healer, all our burden lighten;
Give us peace, Thine own sweet peace, we pray!
Keep us near Thee till the morn shall brighten,
And all the mists and shadows flee away!
—Author Unknown

He knows, He feels, He hears my prayer.
He holds, He lifts me from despair.
Oh, Gentle Healer, touch thou my heart;
make broken pieces Thy work of art.

> WHICH HOPE WE HAVE AS AN ANCHOR OF THE SOUL,
> both sure and stedfast.
> *Hebrews 6:19*

Hope is more than a wish or desire. The dictionary definition of hope is "to long for with expectation of obtaining; confident expectation of that which has to do with the unseen and the future."

Who needs hope more than the loser? The loss may be a husband, wife, child, job, reputation, home, etc. This world is full of things that can be lost. But what are the things that can't be lost? What is the beacon of hope for one in grief?

God has promised: My grace is sufficient. I am a God of love. I am the good shepherd. I am the rock and fortress. I am the bread of life. I am the shelter in the time of storm. I am with you. I have plans for you. I am the light through the dark valley. I never fail or change.

Can we find hope in the sure promises of the eternal God?

Despair is a greater enemy than any disease. Tie a knot in your rope of hope and hang on!

Lord, help me to remember
that in every circumstance
I face today, I am not alone.

—Author Unknown

> BE STILL, AND KNOW THAT I AM GOD.
> *Psalm 46:10*

The loss of a loved one causes many shifts in relationships. It changes relationships with children, grandchildren, in-laws, and all other close relatives, church brothers and sisters, and friends.

These changes are often felt at family gatherings, church services and other functions, at church carry-in meals and school functions.

What is my response to these changes? Hibernate, stay away from large gatherings, go and sit alone and feel self-conscious, keep my mouth shut, avoid people who hurt me, leave as soon as possible.

What can I do to help relationship shifts? Should I go and mix in, even though I feel like I don't fit in, even though I don't know when to share my feelings? Should I talk about my loss with those who understand and show an interest in what is going on in the lives of others?

There are a lot of choices, and some may work for one situation but not another. To help yourself through this touchy time, it is helpful to spend a lot of "still" time with God and allow Him to help you and others adjust to the shifts in relationships.

Be assured that regardless of where you are, what you are doing, or
what you are going through, God is in all things and
in all ways doing that most loving thing concerning you.

—Roy Lessin

> **MY FATHER IS THE HUSBANDMAN.**
> *John 15:1*

>   I Walked a Mile with Pleasure
>   I walked a mile with Pleasure,
>   She chattered all the way;
>   But left me none the wiser
>   For all she had to say.
>   I walked a mile with Sorrow,
>   And ne'er a word said she;
>   But, oh, the things I learned from her
>   When Sorrow walked with me.
>   —Robert Browning Hamilton

There are some blessings we will never receive unless we are ready to pay the price of pain. But, oh, how our human nature shrinks away from pain.

Why are our lives so much like the grapevine? When left unpruned, it grows lush and beautiful vines with no fruit. When pruned, it hangs heavy with fruit. What is of greater value to the Gardener—lush vines or an abundance of fruit?

The great work of redemption was the fruit of great sorrow and suffering.

> *Remember your history and*
> *you will remember God's mercy.*

Grief Relief

If you want relief from your grief for an hour—
    Take a nap.
If you want relief from your grief for a day—
    Go fishing or shopping.
If you want relief from your grief for a week—
    Take a vacation.
If you want relief from your grief for a year—
    Inherit a fortune.
If you want relief from your grief for a lifetime—
    Reach out to other hurting people!
(A revision of the Chinese Proverb—*If you want happiness*)

This treatment for relief sounds simple, but it is often excruciatingly painful. For when you enter into another's anguish, the wound in your own heart is again slashed open. But, as you keep reaching out and your wound keeps getting opened, you find the pain is a little less, because your own wound has healed more from the inside out after each time it was opened.

The side effects from this treatment for grief relief are not only beneficial and healing, but they also provide many reasons for thanksgiving and praise—something thought impossible when loss is first experienced. You can always find someone with more or different hurts.

Thank you, Lord, for Your immeasurable grace.

> WITH ALL LOWLINESS AND MEEKNESS, WITH LONGSUFFERING, forbearing one another in love.
> *Ephesians 4:2*

This is the vocation to which God has called us.

God's call to us is to be everything that does not come natural! Is it any wonder He has to put us through the fire to refine us? This world is the only furnace that can do the job.

Heaven has no wrongs that need forgiveness, no delays for developing patience, no anxieties for testing faith. If we're going to develop a saintly character, it has to be acquired on this agonizingly painful earth. And I have found it is better not to wonder why some have to endure so much more pain than others!

It is not easy to believe that all circumstances, whether of joy or sorrow, which are permitted to come to a child of God are for the purpose of teaching and maturing him in the fruit of the Spirit.

Is my l-o-n-g time suffering transforming me into a long-suffering Christian? Is this oven of pain enlarging my heart with love and understanding for others? Are my losses making me more humble and submissive and more excited about gaining heaven?

Heaven is a prepared place
for a prepared people.

> O YE, BENEATH LIFE'S CRUSHING LOAD,
> whose forms are bending low,
> Who toil along the climbing way
> with painful steps and slow.
> Look now, for glad and golden hours
> come swiftly on the wing,
> O rest beside the weary road
> and hear the angels sing.
>
> —From "It Came Upon the Midnight Clear"
> by Edmund Sears, 1849

Oh, do the angels still sing at Christmastime—that first anniversary Christmas when you find yourself alone and missing that someone? When you find yourself bent low with grief, do the angels still sing?

Oh, yes, you will still find "There's a song in the air, there's a star in the sky!" (Josiah Gilbert Holland) if you can still love and give. Isn't that what life is about, loving and giving and going on with living?

» Perhaps it's time to rest beside the weary road and unload your burden upon the Lord's capable, strong shoulders.

» Perhaps it's time to use your feeble strength to lift another's head and heart.

» Perhaps it's time to refocus from yourself to others.

» Perhaps it's time to be still and listen for the angels' song.

Think about what the "glad and golden hours" of Christ's advent mean in your life.

To give light to them that sit in darkness and in the shadow of death, to guide our feet into the way of peace.

—Luke 1:79

> WE ARE TROUBLED ON EVERY SIDE, YET NOT DISTRESSED;
> we are perplexed, but not in despair; persecuted,
> but not forsaken; cast down, but not destroyed.
> *2 Corinthians 4:8, 9*

These are the words of a victor, a winner, a champion! This is definitely not what you feel like when in the throes of grief. What gave Paul such courage, such keeping-on strength?

Surely it was his focus as expressed in verse seven. *But we have this treasure* [light, knowledge, and glory of God] *in earthen vessels, that the excellency of the power may be of God, and not of us* (2 Corinthians 4:7).

Grief, loss, hurts, and trials all expose my weak, earthen vessel. It seems like bad news when I find out how vulnerable and weak I am, but that really is the best news. For as I realize my humanity, I grasp for God's divinity. Only He can keep me from falling to pieces.

Truly, it is only during the tough times that God's power can be demonstrated from within me.

My prayer is that I can keep on:

Faint, yet pursuing; weak, yet subduing;

Spent, yet renewing; Christ ever viewing.

> WAIT ON THE LORD: BE OF GOOD COURAGE,
> and he shall strengthen thine heart: wait, I say, on the LORD.
> *Psalm 27:14*

Have you ever had to encourage yourself in the Lord? When we are going through the dark valley of grief, we often have to encourage ourselves. Others are getting on with living and have left us far behind in the valley. Sometimes there seems to be no one to encourage us.

In this verse, David is encouraging himself in the Lord. His stronger self is telling his weaker self to WAIT.

How shall we wait so that God might strengthen us? Wait with prayer. Wait with humility and submission. Wait with hope and expectancy. Wait with service.

We think of the lion as roaring courage. But courage does not always roar; sometimes it is the quiet strength of just continuing every day.

Take courage, dear mourner, you made it through yesterday and today, and God will be there for you tomorrow.

When I stand still and wait in expectation,
God will work.

> SEARCH ME, O GOD, AND KNOW MY HEART:
> try me, and know my thoughts: and see if there be any wicked way in me, and lead me in the way everlasting.
> *Psalm 139:23, 24*

This is a difficult prayer when overcome with grief. How can I ask You to search, know, and try me more when I feel like You have done such a painfully thorough job that there is nothing left?

Yet it is only after I accept Your search that I can expect Your guidance.

My Prayer

Sustain me with Your strength—that I might bear it.
Fill me with Your peace—that I might share it.
Satisfy me with Your courage—that I might dare it.
Surround me with Your love—that I might forgive it.
Enfold me with Your hope—that I might give it.
Saturate me with Your grace—that I might live it.

Thank You, Lord, for being in this with me.

> GRACE AND PEACE BE MULTIPLIED UNTO YOU
> through the knowledge of God, and of Jesus our Lord.
> *2 Peter 1:2*

At every hour of life I need God's grace. But, oh, how much more I need grace in grief! I must find moment-by-moment grace for each step through the dark valley.

Someone has defined grace as the ability to do things God's way. That is always an ability we need in our lives, isn't it? And God has grace available for all our needs.

Pardoning grace—forgiveness for my stumbling and faithless questioning; forgiveness for others' insensitivities.

Renewing grace—daily I need renewal in body, soul, and mind. I also need renewed trust in God's sovereignty.

Strengthening grace—grief makes me weak physically, spiritually, and emotionally. Only God's strength will see me through!

Enlightening grace—this valley is very dark; the light of God's love and His Word will guide me out of the shadow of death.

Consoling grace—comes most powerfully from the One who knew sorrows and rejection personally, and who understands us fully.

> *Grace is but glory begun, and glory is but grace perfected.*
> —Jonathan Edwards

> NOT BY MIGHT, NOR BY POWER, BUT BY MY SPIRIT,
> saith the LORD of hosts. Who art thou,
> O great mountain? . . . thou shalt become a plain.
> *Zechariah 4:6, 7*

I find huge mountains, Lord, in this dark valley! Sometimes there are mountains of fear, anger, insecurity, and distrust. Oh, Lord, You know all the mountains! What shall I do with these mountains?

Just ignore them? I can't ignore them, because they are so real. I can pretend to ignore them, but they will reappear again later.

Shall I refuse to go over them and look for an easy way around? There is no easy way; there is no way around them!

Just give up? That is an option, but, Lord, I could not live with the consequences. I would be eternally damned.

Level them into a plain and move ahead? That sounds good, but so difficult. I shall do it, not by my might nor my power, but by Your Spirit!

Help me to remember that my grief, my dark valley, my mountains are not unique! Yes, my experience is uniquely mine, but millions of others have their own unique battles of life. There is no mountain that God cannot help me level, when I face that mountain with faith in Him! God allows mountains in my life, so that I might grow stronger.

Help me, Lord, to say thank You for the mountains!

> BUT THOU, O LORD, ART A SHIELD FOR ME.
> *Psalm 3:3*

Grief has a definite weakening effect on the body, soul, and spirit. So often, it seems like the Lord is carrying me along because I have no power of my own. At the end of a day, I sometimes wonder how I made it through that day.

I find my shield of faith is very weak also, Lord. I need to depend on Your shield in this dark valley. When I find myself swimming in tears of grief, I need Your shield to protect me from self-pity, from bitterness, from painful words of others, from hibernation, from hopelessness, from grief denial, from depression, from fear of the future.

Oh, I feel so vulnerable with raw, sensitive nerves. I am weak, exhausted, and disorientated. I need protection from all the fiery darts of the wicked one.

I must keep this promise in mind. *The angel of the Lord encampeth round about them that fear him and delivereth them.*

I pray for the presence and the shielding
of your mighty angel, O Lord.

> PRAISE YE THE LORD.
> Blessed is the man that feareth the LORD,
> that delighteth greatly in his commandments.
> *Psalm 112:1*

F-right that leads to flight and confusion.

E-nergy exhaustion.

A-nxiety and impatience.

R-esources wasted.

As soon as I notice the effects of my fears, I must re-focus and start counting my blessings instead of fears.

Praise shrinks fear and increases your face value with a peaceful smile—not easy when your heart is breaking—but possible when you count your blessings. You haven't lost everything.

Try this fear/praise exchange!

» Fear of death—praise for God's new mercies this morning. The One who holds the universe holds life in His hands.

» Fear of how long this pain is going to last—praise for mo mentary respite.

» Fear of failure—praise for God's promise, I can do all things through Christ.

» Fear of financial loss—praise God for my riches in glory by Christ Jesus.

» Fear of the future—praise God for strength, direction, and hope that comes one day at a time, and praise Him for the promise of a heavenly mansion.

Where fear dwells, trust and praise do not stay.

Why do people say what they do?

"You can thank the Lord your daughter is in heaven."

Which child would you like to send up there right now?

"Your husband is released from all earthly care and responsibility. Isn't that wonderful?"

Are you ready to tell God to give your husband a "break" so you can practice being a widow with all the cares and responsibilities?

"You can thank God your baby is out of pain, and you'll likely have another baby sometime."

One child can never replace another that you've lost. Lose one and find out!

"You just need to lean on the Lord and trust Him more."

Who leans harder than the one who has all the props knocked out from under him?

"The grace of God is sufficient to see you through this difficult time."

Yes, indeed, but it would help to have some human patience and understanding shown at the same time.

What makes such true statements so trite? When your pain is so deep and people's understanding is so shallow, there is no bridge to span the gap—except forgiveness!

When you truly love, forgiveness is a delight.

> SHALL HE THAT CONTENDETH WITH THE
> Almighty instruct him?
> *Job 40:1*

A grieving person has countless questions, most of which have no answers. Don't prove yourself a fool by trying to answer them—even if you've been through grief yourself. Every person handles his grief differently.

Asking questions is a way of working through the grief. Remember all the questions Job asked? Instead of answering his questions, God asked him more questions. Simply acknowledging the questions sometimes supplies the answers.

Always remember, God can handle your questions. He is not threatened by questions, they do not shock Him, and He does not feel obligated to answer them.

We humans react exactly the opposite to questions. Our intelligence is threatened by questions. We express shock at hearing questions for which we have no answers. Too often, we feel we must fix the person's pain and, therefore, foolishly try to answer unanswerable questions.

"Silence is golden, and a hug is priceless!"

Sympathy is as oil to the machinery of life.

> **WHY ART THOU CAST DOWN, O MY SOUL?**
> and why art thou disquieted within me? hope thou in God: for I shall yet praise him, who is the health of my countenance, and my God.
> *Psalm 42:11*

Thou seest our weakness, Lord!
Our hearts are known to Thee;
Oh, lift Thou up the sinking hand,
Confirm the feeble knee!

Let us in life and death,
Thy steadfast truth declare,
And publish, with our latest breath,
Thy love and guardian care!

Leave to His sovereign sway
To choose and to command;
So shalt thou wondering, own His way;
How wise, how strong His hand.

Far, far above thy thought
His counsel shall appear,
When fully He the work hath wrought
That caused thy needless fear.

—Paul Gerhardt 1860

> YEA, A SWORD SHALL PIERCE THROUGH
> thy own soul also.
> *Luke 2:35*

Joseph and Mary had taken Jesus into the temple where they met Simeon—a man who knew more about their marvelous Son than they could have imagined.

Simeon blessed the parents, and then spoke directly to Mary, making a prediction about her Son. He was preparing her for the great dark valley she would pass through in later years.

How many times does God try to prepare us for what is ahead, but we fail to recognize it until we are in the valley and looking back?

Simeon knew Mary's mother-heart would be deeply pained by her Son's earthly experience of rejection, denunciation, and condemnation. Her heart would be cut severely by His death on the cross.

Many a grieving mother (wife, father, husband) can understand Simeon's words from the depth of a pain-pierced heart.

How often the love which is the source of the purest joy is also the occasion of the most poignant sorrow!

> FOR THOU WILT LIGHT MY CANDLE:
> the Lord my God
> will enlighten my darkness.
> *Psalm 18:28*

We all perceive our time of darkness in different ways at various times. To one, there is always a light in the darkness, a rainbow after the storm. To another, all is doom and gloom. To another, the ways of God are hard, dark, and mysterious. To another, God's ways are unjust. To another, God is the only light and strength that is available.

It has been said, "God's revelation of himself is suited to man's spiritual capacity."

Well, I have to admit to having all the above feelings at various times through my valley experience. That must be evidence that my spiritual capacity fluctuates vastly.

In the end, I must acquiesce to God's faithful, compassionate constancy that provides light for just one step at a time. There is nothing within me to direct myself.

Underneath all the painful, unanswerable feelings and questions is a God, my God, who does not change and who knows it all. He is the light on my step!

Walk in the light and thou shalt see thy path, though thorny, bright;
For God, by grace, shall dwell in thee, and
God himself is light.
—Bernard Barton, 1784-1849

Psalm 6 has been called A Penitential Psalm.

It is saturated with David's grief over his sin. Let's read it now as A Psalm of Grief in the Valley of the Shadow of Death (a paraphrase).

O Lord, rebuke me not in Your anger, neither discipline and chasten me in Your hot displeasure. Have mercy on me, because I am weak, fading, and withering away as a flower. My heart, soul, body, and mind are greatly disturbed. How long will I suffer this immeasurable pain of my loss? Return, O Lord—You seem so far away. Release my soul from this mourning; oh salvage me for the sake of Your compassion. For in the valley of the shadow of death there is only painful silence and I find it so difficult to give You thanks. I am weary with my groaning; every night I drench my bed in tears, and during the day, the tears run freely. My eyes are dull, red, and swollen because of my grief. My heart feels old because of fear of the enemy; anger, guilt, and sorrowfulness surround me. Depart from me, oh enemies—fear, anger, guilt and sorrow! For the Lord has heard my weeping, and He has understood my tears. The Lord has heard and received my prayers, and He will deliver me in His time. He will make me a trophy of His grace; and He will make those ashamed who thought I would never return to "normal."

Indeed, it is a "new normal," but I shall move on beyond my grief valley!

One Day At A Time

One day at a time with its failures and fears,
With its hurts and mistakes, with its weakness and tears,
With its portion of pain and its burden of care;
One day at a time we meet and must bear.

One day at a time to be patient and strong;
To be calm under trial and sweet under wrong;
Then its toiling shall pass, and its sorrow shall cease;
It shall darken and die, and the night shall bring peace.

One day at a time—but the day is so long,
And the heart is not brave and the soul is not strong.
O Thou pitiful Christ, be Thou near all the way,
Give courage and patience and strength for the day.

Swift cometh His answer, so clear and so sweet;
"Yes, I will be with thee, thy troubles to meet;
I will not forget thee, nor fail thee, nor grieve;
I will not forsake thee; I never will leave."

One day at a time, and the day is His day;
He hath numbered its hours, though they haste or delay;
His grace is sufficient; we walk not alone;
As the day, so the strength that He giveth His own.
—Annie Johnson Flint, 1866-1932

Count that day lost, whose low descending sun views
from thy hand no worthy action done.
—Stanford

> **FORGETTING THOSE THINGS WHICH ARE BEHIND,**
> and reaching forth unto those things which are before, I press
> toward the mark for the prize of the
> high calling of God in Christ Jesus.
> *Philippians 3:13,14*

Why do some find such a dreadful thing as grief difficult to leave behind, and why do others simply seem to ignore it? There are those people who pretend never to experience their grief. They jump up from a knockout blow and continue running. Some day, they'll meet their grief head-on.

Others lie down in their grief and find it difficult to rise again to the challenges of life. Others may struggle to get up and go on because they may be plagued with a fear that they'll forget their loved one.

When Paul says, "Forgetting those things which are behind," he surely must mean: Don't allow them to control your future life and hinder you from pressing on. Forgetting traumatic experiences in life is impossible for most of us.

You will never forget the loss of your loved one, and you will many times experience a return of the pain, but as you "press on" you will again learn to laugh instead of cry, sing instead of moan, become involved instead of hibernating.

> *There is a noble forgetfulness—*
> *that which does not remember injuries.*
> —Simmons

> THE LORD IS MY STRENGTH AND SONG,
> and he is become my salvation: he is my God, and I will prepare
> him an habitation; my father's God, and I will exalt him.
> *Exodus 15:2*

This verse is from the song of Moses expressing his thanksgiving for Israel's deliverance from Pharaoh's army. Notice the personal pronouns in the verse. The closer our relationship is with God, the more personal is our expression of praise and thanksgiving.

My grief experience is between me and my God. Each family member's grief experience is between him and his God. As each individual works through his grief, he needs to come to the time when he can personally prepare his heart as a peaceful habitation for his God. How can we best do this? By finding things, people, and experiences to thank and praise Him for—usually a difficult thing to do while the heart is throbbing in grief.

I will sing the grace-filled story,
Of the Christ who walked with me
Through the dark and fearful valley,
Filled with hopeless agony.
Yes, I'll sing of faith's reliance
On my great and sovereign God;
Yield my heart in full compliance
To the path that He has plod.

—Faythelma Bechtel

> THE HAND OF THE LORD IS GONE OUT AGAINST ME.
> *Ruth 1:13*

Naomi had moved to Moab with her husband and two sons, seeking a better life away from the famine in their homeland. After ten years, she found herself ALONE. Her husband and sons had died. Why?

I'm sure Naomi asked the "why" question, and it seems she came to her own conclusion. At the moment of decision she was without hope for herself and her daughters-in-law. Knowing she would never have a husband or more sons, she beggged her beloved daughters-in-law to go back home while she returned to her homeland.

Grief is a time of feeling VERY ALONE, but how wonderful when you find someone who cares and understands! Ruth needed her mother-in-law, and she knew her mother-in-law needed her.

Naomi had proven her loyal devotion to her God in an idolatrous country. Following her example, Ruth declared her allegiance to Naomi and to her God in Naomi's deepest, darkest time of hopelessness. Ruth's own broken heart could readily empathize with the losses of her mother-in-law. Her love would not allow her to leave one so dear to her!

It is a beautiful necessity of our
nature to love someone.

> CALL ME NOT NAOMI (PLEASANT), call me Mara (bitter): for the almighty hath dealt very bitterly with me. I went out full, and the LORD hath brought me home again empty.
> *Ruth 1:20, 21*

A homecoming is an emotional time, especially for losers. Being back home conjures up enormous reminders of the past. Naomi had come home a loser, yet she maintained that God had brought her where she was. Life was bitter, life was empty, but she would wait on God to fill it again. She brought with her a face saddened and worn from painful experiences, but her heart and character were noble and full of love.

"Is this Naomi?" they asked. Yes, indeed, the same woman, but at the same time a very different woman. Grief has a way of softening the sharp edges of our character.

Someone has said,

"Thy will is sweetest to me when

It triumphs at my cost."

What a difficult song to sing, and still more difficult to live. Naomi went home and lived not only her life in God's will, but she also helped Ruth find her place in God's will!

When you can reach out to others, you know you are healing!

He Gathers Every Teardrop

Regardless of the circumstance,
Regardless of the fear,
Regardless of the pain we bear,
Regardless of the tear,

Our God is ever in control,
Performing as He should,
And He has promised in His Word
To work things for our good.

But as a loving Father would,
He sometimes lets us cry
To cleanse the hurt out of our heart,
To wash it from our eye.

Yet gently gathers He the tears
Within His hands to stay
Until He turns them into pearls
And gives them back someday.
—Glenda Fulton Davis, 1980
Used by permission

Oh, Lord, I shall be a very prosperous soul
when all my tears come back as pearls!

> BLESSED BE GOD . . . THE FATHER OF MERCIES,
> and the God of all comfort.
> *2 Corinthians 1:3*

God is the source of all mercy; it is an attribute of God. He is
full of compassion,
  and gracious,
    longsuffering,
      and plenteous in mercy and truth (Psalm 86:15).

This is the character of my God—a bit hard to conceive, much less feel, when I am first thrust into that dark valley experience. When grief dashes my hope, challenges my faith, and smashes my heart, where can I go?

Where can I go, but to the Lord! In time, the love of God shines through the black abyss and renews my hope. In time, the grace of God glimmers through the long night and restores my faith. In time, the longsuffering, mercy and truth of God flickers through the deep darkness of sorrow and begins healing my heart.

I can more easily identify with His compassion because He chose to suffer for me while on earth. He is *the God of all comfort,* who is *touched with the feeling of our infirmities* (Hebrews 4:15).

Only sorrow can speak to sorrow.

> I BEAR IN MY BODY THE MARKS OF THE LORD JESUS.
> *Galatians 6:17*

Life's trials are so difficult;
Pushing in like angry waves of the sea.
I don't understand! "Lord, why me?"

I claimed you for my own, my child;
In the dark Garden of Gethsemane
I never questioned my Father, "Why me?"

The wounds of Jesus are a testimony of His great love for me and His accomplishment of my salvation.

As I emerge from my valley experience, what testimony will my wounds declare? Will I prove myself to be a more understanding and more sensitive person? Will I emerge a wiser, more loving person? Will my relationship with God have greater dimensions?

Oh, God, may my life be a display of Your tender nourishing and healing of my troubled, broken heart!

> THE GOD OF ALL COMFORT;
> who comforteth us in all our tribulation,
> that we may be able to comfort them which are in any trouble,
> by the comfort wherewith we ourselves are comforted of God.
> *2 Corinthians 1:3, 4*

It's amazing how the valley experience displays both the character of God and my limitations! "He knows our frame; he remembers that we are dust." "In all our afflictions, he is afflicted." Cannot I trust His character? He will not desert me! He will not reject me! He suffers with me!

In the valley, I learn not to trust my feelings, my emotions, my pain, or my reasoning ability. In the valley, I must learn to trust the character of God, the sovereignty of God, the judgments of God.

As I submit, commit, and trust, help me, Lord, to be endowed with the power and comfort of Your love so that I may comfort those around me.

Have I found fellowship "with Christ in His sufferings"? Have I found comfort in His grace and love?

It is in receiving that we can give and
it is in giving that we receive.
—St. Francis of Assisi

> IS ANY AMONG YOU AFFLICTED? LET HIM PRAY.
> *James 5:13*

Afflicted means troubled, distressed, upset, bothered, miserable, worried, anxious, disturbed, and many more distraught emotions. Life is at times full of stressful feelings, but when one is grieving, all such feelings are exaggerated many times over.

It is Satan's delight to try to drown us in self-pity. And that isn't a difficult chore for him when we focus on our feelings. Sad to say, it sometimes feels self-satisfying to be the victim. Once you're down, you might as well stay down. It almost feels good to feel bad! It is a strange paradox, but it is so true when we don't focus on God and His goodness.

God does not want us to be a victim, but a victor! God in His grace and mercy knows just how much I can take. He has promised that I will pass through the waters and not go under. He has promised peace in pain. He has promised light in darkness. He has promised bread in famine.

What is my part in this affliction? Focus on God and pray!

I will pray for strength in sorrow, mercy in misery,
assurance in anxiety.

> BEHOLD, I STAND AT THE DOOR, AND KNOCK:
> if any man hear my voice, and open the door, I will come in to him, and will sup with him, and he with me.
> *Revelation 3:20*

A door can be a passageway or a barrier. Oh, how I remember that closed-door feeling—it was a barrier, not a passageway. The Lord had taken my loved one and shut the door! I was shut away from God and my loved one. I was shut away from normal living and normal people. I was closed up in a dark, lonely closet of grief and pain.

What I failed to realize was that the knob was on the inside! I was in control of that door. Christ was on the outside knocking, waiting to come in and comfort me. He was waiting to come in and build a better relationship with me. He was waiting to fellowship and to sup with me.

But I didn't open the door, because I had nothing tantalizing and nourishing to offer Him. I was empty and hurting beyond words.

Finally, in desperation and almost hopelessness, I went to the door and offered Him my broken and contrite heart. Eagerly He accepted my offering and came in. It was just what He wanted. And He supped with me, and I am still supping (partaking of the nourishing food He provides!) and learning from Him.

A door is either a passageway or a barrier.
The door to God is always open.
The door to my heart I must open.

> THE LORD IS GOOD UNTO THEM THAT WAIT FOR HIM, to the soul that seeketh him. It is good that a man should both hope and quietly wait for the salvation of the LORD.
> *Lamentations 3:25, 26*

Wait in the darkness,
Wait in the pain;
Wait in the deep valley,
Wait in the tears of the rain.

Wait in your sorrow,
Wait in your fear;
Wait when hope seems all gone,
Wait in faith—Jesus is near!

Wait 'til there comes light,
Wait 'til there's peace;
Wait for God's blessings of
Grace, mercy, and sweet release.

Release from the fear,
Release from the dread;
Release from the darkness,
Release to trust all He's said.
—Faythelma Bechtel

Waiting is a test of trust, courage, and patience.
Wait with prayer, with humility, and with expectancy!

> BUT THOUGH HE CAUSE GRIEF,
> yet will he have compassion according
> to the multitude of his mercies.
> *Lamentations 3:32*

I saw from the window the bare-branched maple bejeweled in the gray of a beginning day. Droplets of rain hung like hundreds of shimmering pearls in the gloomy dawn.

I bowed my head and prayed, "Oh, Lord, make my grieving tears of more value than these many pearls. May they brighten this dark path and become stepping stones to You."

Then, transfixed, I watched as the sun's early beams sent out bright rays that transformed the pearls into sparkling diamonds. Diamonds reflecting the sun's rays, creating brilliant flashes of red, green, blue, and gold.

Another prayer came to my lips. "Please, God, may the grace of Your Son's rays disperse this gloom and turn my tears to diamonds. Diamonds that flash and sparkle with Your love and grace, giving hope to other hurting souls, displaying Your transforming, healing power! Amen."

The Christian graces are like perfumes, the more they are pressed,
the sweeter they smell; like stars that shine brightest in the
darkest skies;
like trees which, the more they are shaken, the deeper root they take,
and the more fruit they produce.

—Beaumont

> TURN UNTO THE LORD YOUR GOD:
> for he is gracious and merciful, slow to anger, and of great kindness, and repenteth him of the evil.
> *Joel 2:13*

It has been said that sometimes God subtracts (allows subtractions) in order to add.

My first reaction to that statement was, "Subtracts, indeed! He takes away our daughter and my husband, and then adds pain, sorrow, and an overload of work and worry for me to carry."

For months and months, it seemed God truly was adding to my life nothing but painful burdens. It took a long while before I realized it was my attitude toward everything that had happened which really increased the stress load.

Later, I began to understand God's refining purpose. God's ultimate goal in my life is to bring me into a closer relationship with Himself, a real relationship that will glorify Him and totally fill my every need.

Living in a sinful world subtracts from our lives, leaving us in painful, seemingly hopeless situations. But God is always there in the shadows waiting to change these subtractions into additions and multiplications.

Misfortune is not forever mournful to the soul that accepts it; for such do always see that in every cloud is an angel's face.

—Jerome

> BLESSED IS THE MAN WHOSE STRENGTH IS IN THEE;
> in whose heart are the ways of them. Who passing through the valley of Baca (weeping) make it a well;
> the rain also filleth the pools.
> *Psalm 84:5, 6*

Are you asking along with me, "Lord, however can I turn this valley of weeping into a well—a well of refreshing water"?

Only the reign of God's love in this gloomy valley and the reign of His grace in my heart can transform these bitter tears into sweet, healing waters. Help me, Lord!

While God promises the comforting rainbow of His understanding, compassion, and mercy, He also places before my feet some stepping stones recorded for us in Psalm 37:1-7.

Fret not!
    Trust in the Lord!
        Commit thy way!
            Delight thyself in the Lord!
            Rest in the Lord.
                Be still and know God!
            Do good!

Please, keep my feet on these stepping stones, and then I can be assured of the refreshing showers of Your love and grace for each trying day.

Thank You, Lord!

> HE THAT DOETH THE WILL OF GOD ABIDETH FOR EVER.
> *1 John 2:17*

In Me

Whatever Thy will may be,
Lord, let it be done in me.

For now thou hast laid me low
In the mystery of woe;
I am shut from speech and song,
I am weak who was so strong,
And my soul is known of Thee
In her great adversity.
So I clasp Thy feet, and say,
"Thou hast given and taken away."

I know that I serve Thee so,
In lying all meek and low;
I know Thou wilt deeply bless
In trusting and quietness;
And I cry through the gracious gloom
Of the fig tree's perished bloom,
"Whatever Thy will may be,
Lord, let it be done in me."
—E. E. Hickey, 1890

Surrender to God's will is never a risk.

> CHASTENING IS GRIEVOUS,
> but afterward it yieldeth the peaceable fruit of righteousness
> unto them which are exercised thereby.
> *Hebrews 12:11*

Someone has said, "Faith discerns the good hand of God in all things."

When I truly love God, I will endeavor to trace God's loving hand through my pain and sorrows. When the grace of God keeps me plodding on, with the light of God on each step, I will sense enablement by the power of God and strength from the love of God. Then I will successfully pass through the valley of the shadow of death by the goodness of God.

Thank You, God, for Your good hand that brings forth the peaceable fruit of righteousness. Help me as I exercise in accepting all things from You for my good!

> *'Tis sorrow builds the shining ladder up,*
> *Whose golden rounds are our calamities,*
> *Whereon our firm feet planting, nearer God*
> *The spirit climbs, and hath its eyes unsealed.*
> —James Russell Lowell, 1819-1891

> For I am the Lord, I change not; therefore . . .
> we are not consumed.
> *Malachi 3:6*

The following story is told of an elderly, retired, wheelchair-bound music professor who lived in a boarding house. A neighbor boarder would ask him the same question each morning and receive the same answer each morning.

"What's the good news, friend?" came the question.

The old man would pick up his tuning fork and tap it on the side of his wheelchair. Then he would reply, "That's middle C! It was middle C yesterday; it will be middle C tomorrow; it will be middle C a thousand years from now!"

When my life has been changed by a major loss, middle C is exactly what I need. I find myself feeling very confused, apprehensive, and insecure.

I need the unchangeable middle C—Christ! He is the same yesterday, today, and forever. He is the rock I can cling to, the shelter I can take cover in, and He is the anchor I can trust to hold.

*Because He is unchanging, I will not
be consumed by my grief or by Him!*

> WHEREFORE THE KING SAID UNTO ME,
> Why is thy countenance sad?
> *Nehemiah 2:2*

The king gave Nehemiah a golden moment, a time to express his feelings honestly. How often does God give me those golden moments and I fail to see them because I want to hide my true feelings?

Someone has said, "Golden moments in the stream of life rush past us, and we see nothing but sadness; the angels come to visit us, and we only know them when they are gone."

Before Nehemiah answered the king, he sent a telegraph to God, *So I prayed to the God of heaven* (Nehemiah 2:4). Then he presented his cause wisely and honestly and obtained the king's sympathy and help.

Lord, I have prayed and prayed, and now I come to You with my honest answer and plea. My heart is shattered. Please put it back together in a useable fashion—with less pain and more beauty. Repair my relationship with You and others. And please, heal these raw, agonizing emotions.

Of all the sad words of tongue and pen,
the saddest are these: It might have been.
—Whittier

Help me, Lord, never to live in the world of it-might-have-been.
For with You all things are possible—even a healed, joyful heart.

> **BEHOLD I AM LAYING IN ZION A STONE**
> that will make men stumble, a Rock that will make them fall;
> but he who believes in Him [who adheres to, trusts in and relies
> on Him] shall not be put to shame nor be
> disappointed in his expectations.
> *Romans 9:33, Amplified*

Jesus is that Rock. In my life, will He be a stumbling stone, a crushing stone, or a stepping stone? It is so easy to stumble in the dark valley. It is also easy to feel crushed and hopeless.

I like the Amplified's expansion on the word "believe." It says, *(who adheres to, trusts in, and relies on Him!)* That's my duty! Adhere like Super Glue, trust as a baby in his mother's arms, and rely as a bird does on his wings.

Then comes God's unfailing promise, *I shall not be put to shame nor be disappointed in* [my] *expectations.*

Well, Lord, I don't want to be ashamed of myself or shameful to You in this grief experience, so show Yourself strong and uphold me on all my leaning sides! Make me a token of Your grace and a story of Your love and mercy, so that neither my expectations nor Yours will be a disappointment!

We trust as we love, and where we love. If we love Christ much, surely we shall trust him much.

—Brooks

> THE LORD IS MY SHEPHERD, I SHALL NOT WANT.
> *Psalm 23:1*

What you have in your Good Shepherd
is greater than what you have lost in life.

Have I learned with David, *"I shall not want"?* I'm still learning!

One meaning of "want" is that I shall have no lack of proper care or management. A broader meaning, which David implies, is that I am utterly contented in the Good Shepherd's care and desire nothing else.

After I have loved and lost, I am tempted to question my Shepherd's proper care and doubt His management. The more I focus on what I have lost, the more discontented and unhappy I will become.

Because I Loved
Because I have loved and lost;
Pain and heartache are the cost.
I accept the price and hold love dear,
For blessings You sent year after year.
My loss is but heaven's gain.
My dear Shepherd will sustain.

I thank You, dear Shepherd, for Your care and management.

Thank You for being greater than what I have lost.

> **KEEP NOT THOU SILENCE, O GOD:**
> hold not thy peace, and be not still, O God.
> For, lo, thine enemies make a tumult:
> and they that hate thee have lifted up the head.
> *Psalm 83:1, 2*

The enemies in the valley are a bit different from the enemies David is here referring to, but they are no less enemies of God and of you and me.

Grief's valley enemies are many! Fear, blame, anger, resentment, exhaustion, hopelessness, loss of normalcy, loss of identity, loss of direction, immobilization, sickness, hibernation, and many unidentified emotions, just to name a few.

You may feel like a BLOB surrounded by flying bullets. What do these enemies want with you? They want to ruin your relationship with your Good Shepherd. They want to cast your soul into hell.

Lift up your head! You are not a BLOB, you are a child of God. God is not silent; He is simply waiting for you to listen. He is waiting for you to believe that He loves you and is taking care of you.

He is waiting for your submission and admission, *That men may know that thou, whose name alone is JEHOVAH, art the most high over all the earth (*Psalm 83:18).

> *To do or not to do, to accept or not to accept, I leave to thee;*
> *thy only will be done in me; all my requests are lost in one,*
> *Father, thy will be done!*
> —Charles Wesley

> A MAN THAT HATH FRIENDS MUST SHEW HIMSELF FRIENDLY;
> and there is a friend that sticketh closer than a brother.
> *Proverbs 18:24*

Thank God for friends who stop and say they care.
Thank God for those who say, "You're in my prayer."
Thank God for callers—just checking in
To find you okay, or in a tailspin.
Thank God for friends who invite you for lunch,
And for the indifferent—without a hunch.
Thank God for those with time for a hug.
Even be thankful for those who just shrug.

Friends can never replace God, endure as patiently as God, or understand as clearly and accurately as God. But they surely can supply a human connection when you so badly need one.
—Faythelma Bechtel

Oh, the comfort—the inexpressible comfort
of feeling safe with a person.
Having neither to weigh thoughts,
Nor measure words—but pouring them
All right out—just as they are
Chaff and grain together—
Certain that a faithful hand will
Take and sift them—
Keep what is worth keeping—
And with a breath of kindness
Blow the rest away.
—Dinah Maria Mulock Craik, 1826-1887

Friendship improves happiness, and abates misery,
by doubling our joy and dividing our grief.
—Addison

> I AM GOD, AND THERE IS NONE LIKE ME . . .
> My counsel shall stand, and I will do all my pleasure.
> *Isaiah 46:9, 10*

Someone has said, "My disappointments are God's appointments."

That's a rather difficult one when His appointment is to meet me at a graveside and later in the valley of the shadow of death. My human heart and mind reasons—*surely, God, there would be a better place to meet. I'm deeply disappointed in the direction my life has taken. Things certainly aren't going according to my plans, Lord. This has got to be the most heart-wrenching place to meet.*

But God sees beyond today. God has eternal plans for me, and what He decides is best for me will truly be the best when I accept His plans and acknowledge His counsel.

I am living in the painful, temporary NOW, while He is planning for the glorious, eternal FUTURE!

> *Open my eyes that I may see,*
> *Glimpses of truth Thou hast for me.*
> *Silently now I wait for Thee,*
> *Ready, my God, Thy will to see.*
> —Clara H. Scott, 1841-1887

> *God permits human life to be honored as the theater*
> *in which the great tragedy of conflict*
> *between evil and good is displayed.*

> EVERY GOOD GIFT
> and every perfect gift is from above…
> *James 1:17*

We have been deliberately chosen, singled out, selected by our loving Lord, appointed by God for our new status—widowhood (widower, childless).

The Lord, who had given me singleness and marriage as gifts of His love, had now given me this new gift of widowhood. Would I receive it from His hand? Would I thank Him for it? (Elizabeth Elliot)

"Although it is very difficult, in time we must accept our gift, thank God for it, and reach out for our new assignment.

"Open hands should characterize the soul's attitude toward God—open to receive what He wants to give, open to give back what He wants to take. Acceptance of God's will means relinquishment of our own will. If our hands are full of our own plans, we cannot receive His plans.

"The Lord is gentle, full of compassion, and patient. He tenderly sees us through our initial grief and shepherds us into His plans for our future."

—Leona Choy, used by permission

Deserted—something I never prayed for,
but something I must pray through.
God is my hope and He never deserts me!

> HE LEADETH ME BESIDE THE STILL WATERS.
> *Psalm 23:2*

Still waters—sounds so peaceful, so calm, so undisturbed—so opposite from the way I often feel while working through my grief. And oh, how kindly He leads me. He doesn't push, pull, or force my nose into the still waters. Sometimes I feel as though some people try to push or pull me out of my valley experience.

Thank You, Good Shepherd, for leading. You aren't rushing far ahead of me calling, "Hurry, catch up! You're going too slowly." You are right beside me, leading me, and sometimes carrying me. You know what I need at just the moment I need it.

You gently lead me to the still waters of Your Word where I find refreshment in Your company, in Your compassion, and in Your concern.

The still waters flowing beneath Your throne of grace are effervescing, full of mercy and grace to help in my time of need.

I must not worry about tomorrow. You will provide light and strength for just one step at a time.

It is one thing to know about God,
and another thing to know God.

> IF A MAN DIE, SHALL HE LIVE AGAIN?
> all the days of my appointed time will I wait,
> till my change come.
> *Job 14:14*

Surely Job felt he was only a "heartbeat from heaven." The losses in his life had zoomed in to focus on one subject—DEATH.

Death: It is sure, it is coming, and it is terrible! Is this the way you and I view death? Does death leave us in the valley of fear without hope? Certainly not!

Someone wrote the following acrostic about death. When we look at death in such a positive light, we can surely find comfort and peace for our hurting hearts.

D—Deliverance from pain, heartache, fear, and evil.

E—Eternal life is waiting for us and our loved ones.

A—Assurance of God's love and care.

T—Tears bringing release and cleansing.

H—Heavenly home and hope of seeing my Lord and loved ones again.

Death is life's greatest victory for the Christian.

> BLESSED ARE THE MERCIFUL:
> for they shall obtain mercy.
> *Matthew 5:7*

Have you been seeking mercy in your distress? We know and appreciate God's mercy. But it feels so good when someone with human skin expresses compassionate mercy toward me—especially when my emotions are weak and my thoughts unbalanced.

It is all right to long for mercy, but beware when you forget the needs for mercy of those about you. Others are also hurting and needing mercy. I must make a deliberate effort to feel with others. I must be kindly in my judgments. I must be slow to condemn and quick to commend. Mercy gives and serves. It is in being merciful that you will receive mercy.

Sometimes being merciful hurts too much. The closer you get to another's pain, the more it becomes yours. Too much sympathy, too much pity, may disable you both.

Remember, you are not there to be suctioned into the vortex of the other's pain—that renders you helpless. You are there to offer that rope of hope that allows him to be pulled up onto firm ground.

Your empathy and understanding mercy must acknowledge the right and the wrong attitudes in dealing with grief.

> Nor is there any mediator between us,
> Who may lay his hand upon us both.
> *Job 9:33 (NKJV)*
> For there is one God, and one mediator between
> God and men, the man Christ Jesus.
> *1 Timothy 2:5*

Job wanted someone to mediate between him and God, and, no doubt, between him and his friends. He had opened his pain-torn heart to his friends only to have it slashed open some more. He longed to communicate with someone who could truly know and understand his thoughts, with someone who cared enough to prove it by being merciful.

When I think of Job's dilemma, I am extremely thankful for my Mediator! Yes, I can open my heart to caring friends. They may sympathize, offer help, pray, and feel with me, but they may also go away drawing their own conclusions, adding their own narrow-minded answers to my situation.

Christ, as my mediator, knows my thoughts before I think them. He understands everything that has contributed to making my personality and creating my emotions. He understands why my grief affects me the way it does. He knows the longings of my heart. He hears my prayers before I speak them. He knows the limitations of my humanity.

Thank God I have such a Mediator!

> BEHOLD, WE COUNT THEM HAPPY WHICH ENDURE.
> Ye have heard of the patience of Job, and have seen the end of
> the Lord; that the Lord is very pitiful, and of tender mercy.
> *James 5:11*

God's view of suffering and man's view are often at the opposite ends of the spectrum.

As humans, we do not enjoy or want suffering, and I'm sure God does not enjoy seeing us suffer, but He alone knows the measureless value and results.

We like to try to remove or explain away our suffering and even the suffering of others. It is dangerous to attempt to be an "amateur providence" in anyone's life. It is never my business to set myself up as a deliverer from grief or suffering. That is idolatry, and it will lure the sufferer away from God.

Consider Job, who had no idea about God and Satan's discussion concerning him. All he knew was that he hurt, he was tired of life and his miserable comforters, and he desired an audience with God. He did not enjoy his suffering, he could not explain his suffering, and he could not satisfy his friends.

Always remember, we are not in on all of
God's secrets, so just keep trusting!

> WHERE IS GOD MY MAKER,
> who giveth songs in the night?
> *Job 35:10*

There is a story of a little bird that never sings the song its owner wants to hear while its cage is full of light. It may sing a note or several notes, but will never sing the entire song until its cage is covered and the sunlight is shut out.

Is there a lesson here for me, Lord? When the night is so black we wish for one star, when the compass of hope is gone, when it seems heaven refuses every desperate prayer, is that the time for a song?

Why is it so difficult for our human hearts to know the richness of God's love and its comforting, satisfying completeness while we're living in the sunlight? It takes the dark, storm-threatening skies put side by side with His all-sufficiency to bring the song of deliverance to my lips!

Help me, Lord, to learn to sing the complete song of Your mercy, love, and grace in the dark! Instead of wondering where You are, help me to sing, "God is waiting in the stillness." Instead of dreading the morrow, help me to sing, "Great is Thy faithfulness." Instead of seeking answers for the "why" questions, help me to sing, "I surrender all."

These are the songs You are waiting for me to sing.

> It is he that sitteth upon the circle of the earth,
> and the inhabitants thereof are as grasshoppers.
> *Isaiah 40:22*

Can you imagine our great God sitting upon the circle of the earth so high above us, looking down on His creation who appear as minute grasshoppers? Yet, that all-knowing God is controlling the number of days for each of His children. And He is one-by-one separating and calling ones from their earthly family circle to come to heaven and enter their heavenly family circle. Who are we to question His choice and doubt His wisdom?

Unbroken Family Circle
Our family's love circle is broken,
But let us not despair,
For she is living in the greatness
Of God's love over there.

She has joined our new family circle
Far on the glory side,
Let us each love, lift, and labor on,
'Til with her we abide.

Let's each keep that circle unbroken
By giving God our all,
Making sure our mansion in heaven—
Faithful until His call.
—Faythelma Bechtel

> HE GIVETH POWER TO THE FAINT; AND TO THEM THAT
> have no might he increaseth strength.
> *Isaiah 40:29*

Grief and sorrow make you faint, weak, and fragile; and the more we focus on our weakness the greater it increases.

It is time to change my focus as soon as I begin feeling too weak and apathetic to move ahead. God's strength is waiting to replace my weakness.

God knows the effects of grief and sorrow; therefore, He promises power and strength!

God has three R's for overcoming our weakness:

Remain—faithful and wait for His direction

   Rely—upon him and trust in His promises.

     Roll—your weakness and fears on Him.

God is faithful who has promised. When I remain, rely, and roll, I will experience God's power and strength.

God does not call the qualified, He qualifies the called. When God calls me to suffer, I am tempted to wonder what qualifications I could possibly need. Does it take qualifications to suffer? Is the call to suffer a mission I can complete for His glory?

Can I pray, "Lord, qualify me to suffer for Your glory"?

> I AM THE GOOD SHEPHERD, AND KNOW MY SHEEP.
> *John 19:14*

Can I explicitly trust the Good Shepherd to lead me and control my life? Yes, I can! Do I explicitly trust my Good Shepherd? The answer to that question depends on my view of Him. Too often, I view Him with my human eyes rather than my spiritual eyes. I think of Him with the human reasoning rather than my spiritual discernment. When I do this, my Shepherd becomes the object of doubting, questioning, and distrust.

Oh, Lord, open my spiritual eyes and direct my mind from the hopeless impossibilities in my life to the absolute assurance of Your power to lead me as only a Good Shepherd can.

I belong to You because You deliberately chose me and paid for my redemption with Your blood. Surely I can trust that kind of love.

Now when life has suddenly brought me to the valley of the shadow of death, You will not leave me in the darkness to find my way alone. You are the ever-faithful God, the ever-trustworthy Shepherd, the ever-patient and understanding Master.

May I always remember throughout life—whatever may come—that there is no substitute for a keen awareness of my Shepherd's compassionate presence and direction.

Help me continually to sense Your calming, reassuring presence.

> LOOKING UNTO JESUS THE AUTHOR AND
> finisher of our faith.
> *Hebrews 12:2*

When I begin to feel panic, unrest, and insecurity, I know it's time to check on my focus. It is indeed unsettling to focus on the unwanted, unplanned for, and undreamed of changes that sometimes come into one's life. When I focus on my abnormal present life and uncertain future, I panic!

A wrong focus becomes like a huge mountain. I cannot scale it; I cannot walk around it. I look for an opening, an out—but there is none until I look to Jesus. He is the beginner and finisher, the leader and the upholder of my faith.

The One who holds the universe in His hands can surely assist me with my mountain.

It requires a definite effort on my part to change my focus from my fear, my confusion, my pain, and my doubt and to focus on Jesus. I will focus on His unchanging character, His eternally true promises, and His ceaseless love and care for me.

What He has begun in me, He WILL finish for His glory.

> FOR HE *(GOD)* HIMSELF HAS SAID,
> I will not in any way fail you nor give you up nor leave you without support. *(I will)* not, *(I will)* not, *(I will)* not in any degree leave you helpless, nor forsake nor let *(you)* down, *(relax My hold on you)!* [Assuredly not!] So we take comfort and are encouraged and confidently and boldly say, The Lord is my Helper, I will not be seized with alarm—I will not fear or dread or be terrified. What can man do to me?
> *Hebrews 13:5, 6, Amplified*

You Will Not Be Abandoned!
Thank You, Lord, for Your promise!
This is a promise to stand on
when I am weak.
This is a promise to enclose my mind in
when I'm insecure.
This is a promise to cling to
when I'm fearful.
This is a promise to enfold my heart in
when I'm overwhelmed with my circumstances.
This is a promise to embrace
when I am lonely.
This is a promise to pack into my mind
when I am bewildered.
I will not be abandoned!

Abide—locate, move in, unpack, and settle down...
Abide in Me, and I will do the same in you.

> AND THE LORD GOD SAID,
> It is not good that the man should
> be alone; I will make him an help meet for him.
> *Genesis 2:18*

God created humankind with a two-fold need: for fellowship with Himself and companionship with others. When the link between a married couple is broken by death, the need for some measure of coupling (interaction) with others is important.

—Leona Choy, used by permission

At times, aloneness encloses me in an almost smothering grasp. The silence of aloneness bounces off the walls. The feeling of aloneness knots my stomach. It is then I must divert my mind to other things of importance or to reaching out to someone lonely or in need.

J. Oswald Sanders was no stranger to grief. He lost two wives, and after 17 years as a widower he advised the lonely: "If they would abandon the search for someone to care for them, and set themselves instead to care for someone else, they would be amazed to discover that their loneliness was quite bearable, even if it was not entirely banished."

So, when you're feeling it is not good for man (or woman) to be alone, call and share with someone a verse that gives you courage. Take a bouquet or fruit basket to someone sick or shut-in. Go visit another person who has lost a loved one.

If you don't enjoy being lonely, build bridges instead of walls.

> WHEN JESUS KNEW THAT HIS HOUR WAS COME, . . . [he loved his own unto the furthest extent . . . he] began to wash the disciples' feet...ye are clean, but not all...If ye know these things, happy are ye if ye do them.
> *John 13:1, 5, 10, 17*

The heart of Jesus was heavy with the knowledge that He would die for the sins of the world. His heart was pained and in deep grief because one whom he had chosen, one whom He loved, one whom He had taught the ways of love, had made a decision to reject and betray the One who wanted to be his personal Lord and Master.

At this exceedingly agonizing hour, Jesus reached out in humble love to each of His disciples. He washed their feet as a willing servant.

Did they understand that He was teaching them to seek servanthood? Did they understand that He was displaying the way of humble love, not pride and self-seeking? Did they understand He was showing them the best way to deal with personal pain and loss is to reach out and touch someone else's life with love? Did they understand the joy in obedience?

> *Oh, Lord, help me to understand what*
> *Your example is meant to teach me!*

> BEHOLD, I HAVE GRAVEN THEE UPON
> the palms of my hands;
> thy walls are continually before me.
> *Isaiah 49:16*

When life's skies are darkened with trial after trial, and when waves of sorrow and pain seem almost to drown the soul, it is not uncommon to cry out, *Hath God forgotten to be gracious?* (Psalm 77:9).

David and Israel often made that cry, or a similar one. God's people throughout the ages have uttered that cry. During the reigns of Stalin and Hitler, scores of suffering prisoners cried out, "If there's a God, why does He allow things like this to happen?"

Has God forgotten to be gracious? No, God does not forget. God is always gracious. I must remember that You call Your sheep by name! You care for me. I am always in Your sight and in Your heart, on Your lips and in Your hand!

There is no circumstance or calamity that can remove me from Your care—only my deliberate rebellion can remove me from that place of quiet rest beneath Your wings divine.

Oh, Lord, assist my aching heart and maintain a right spirit within me that I may continually submit my will to Your control. Help me to remember that, in spite of the trauma of what I am suffering now, You have my best interest in Your heart and plan.

I will, by Your grace, become a better person because of what I am suffering now!

> The Lord hath his way in the whirlwind and in the storm,
> and the clouds are the dust of his feet.
> *Nahum 1:3*
> Hereafter shall ye see the Son of man sitting on the
> right hand of power, and coming in the clouds of heaven.
> *Matthew 26:64*
> And a voice came out of the cloud, saying,
> This is my beloved Son: hear him.
> *Mark 9:7*
> H was taken up; and a cloud received him out of their sight.
> *Acts 1:9*
> Behold, he cometh with the clouds; and every eye shall see him.
> *Revelation 1:7*

Notice how clouds and Christ and His presence and power are connected in the above verses. Have you found His message for you in the clouds?

» Lord, my life is shadowed with clouds; help me always to find You in them.

» Clouds of grief and clouds of deep sorrow;

» Clouds of wondering what will come tomorrow.

» Clouds of fear and clouds of booming thunder;

» Clouds that flash lightning and divide asunder.

» Clouds bring faith, and clouds bring promise of Your reign;

» Clouds bring assurance that You are in my pain.

» Clouds are where Your sovereignty and my frail humanity meet.

» Clouds are a proof of Your presence; clouds are the dust of Your feet.

There is sunshine behind every cloud.

> MASTER, CAREST THOU NOT THAT WE PERISH?
> *Mark 4:39*

All too often, when we are happy, healthy, and heartless, we express shock at other Christians' struggles, trials, sorrows, and grief. We tend to think that God is punishing or disciplining those hurting hearts. We glibly pass our judgment that, if they lived their lives as they should, they would not be having such problems. (Sometimes we do bring hard things on ourselves, but God has a merciful way of dealing with us that others will not understand.)

When I am one of those with a hurting heart, my first thought is like the disciples in the storm-tossed ship, "Don't You care, Lord?"

What we seem to forget is that Christ is in every storm-tossed ship. Just because He's in the ship doesn't mean there won't be any storms! And just because Your life is in a terrible storm now, doesn't mean Jesus doesn't care. The storm may not have anything to do with disciplining you. God disciplines His children; He polishes them. Remember Job—the perfect man whom God polished and proved to Satan that he was made of gold.

When I feel the storm raging and tossing my boat high on the waves of fear, distress, and anguish, oh Lord, help me to remember You are in the boat with me waiting for the right moment to say, "Peace, be still."

The Creator of the storm can surely control it.

> HE SHALL SEE THE TRAVAIL OF HIS SOUL,
> and shall be satisfied.
> *Isaiah 53:11*

It is my duty to pass on to the next generation the knowledge and experience of the goodness, mercy, and grace of God. How will I do it? Will I emerge from my valley experience bitter or better? Have I left behind my anger and learned to walk on in God's assurance? Have I exchanged my fears and frustrations for forgiveness and faith?

The grief, bruising, and cross were the travail of Jesus' soul. That travail was finally satisfied by the death and resurrection of Jesus. That travail was for the redemption of you and me.

Can I look on the travail of my soul and be satisfied? What education from the University of Adversity will I pass on to those who follow me?

No difficulty can arise, no dilemma emerge, and no seeming disaster descend on the life without eventual good coming out of the chaos.
—Phillip Keller

What good? In an on-going way, God's goodness provides for all my needs. God, through Jesus, pardons all my sins and understands all my humanity.

God's grace keeps me under the shadow of His wings.

> CHOOSE YOU THIS DAY WHOM YOU WILL SERVE…
> as for me and my house, we will serve the LORD.
> *Joshua 24:15*

Though we are not asked to choose our trials, we certainly must choose how we are going to deal with them. With each trial we face comes the challenges of different choices.

Do I choose to accept what has changed my life, or do I try to reject it? Do I choose to feel secure in God's love, or do I sever myself from God by carrying bitterness? Do I choose to submit to God's will, or do I stubbornly go my own way? Do I reach out or withdraw? Do I choose to be gentle, or am I grouchy? Do I choose to be forgiving or unforgiving? Do I choose to be joyful for what I have, or do I choose to be jealous of others because of the unfairness of life? Do I feel satisfied that God has a good purpose for my pain, or am I saturated with self-pity?

When I choose to trust in the sovereignty and goodness of God in spite of painful losses and changes, I am making it easier for myself and also for those who follow me.

Yes, hard things come and they pass, but they leave their mark—it's my personal choice to transform the trial into a growth mark and not a bitter rotten spot.

Expressing an active faith in God
always brings positive results!

> STILL HE HOLDETH FAST HIS INTEGRITY,
> although thou movedst me against him,
> to destroy him without cause.
> *Job 2:3*

Integrity is a many-faceted word. Honest, trustworthy, constant, and reliable are only a few synonyms. Looking at integrity in grief further expands and brings some serious challenges to the meaning of the word and to living it out in daily practice.

God was able to tell Satan that Job kept his integrity even after he, Satan, had taken everything from him. Job's integrity had a solid foundation, and God was confident of his servant's faithfulness.

If possessions had added to Job's sense of integrity, they were gone. The integrity of Job's health was gone. Some say that the integrity of Job's wife was gone, but I question that. (She was still with him at the conclusion of his colossal trial, and they had another beautiful family together.) Job's emotional integrity was definitely demolished by the time his friends were finished with him. Their questions and accusations damaged the integrity of his speech. They made him feel that he needed to defend himself—never a good reaction.

What was left of Job's integrity? His faith and trust in the character of God was left. *I know that thou canst do everything* (Job 42:2).

Read that last chapter of Job, and be assured that keeping your integrity is possible in any trial when your trust is in God.

> THOU ANOINTEST MY HEAD WITH OIL.
> *Psalm 23:5*

When sheep become too bossy and begin butting heads, the shepherd puts oil on their heads so they glance off instead of hurting each other.

I must admit to being like a bossy sheep sometimes. I have been wounded and caused wounds by butting heads with others, and even sometimes by butting against God's purposes.

Life is full of wounds! Some wounds I bring on myself, some come unjustly. Illnesses have wounded me deeply. Age can become a daily wound. Disappointments and rejection often wound me severely. And only the Lord knows the depth of my wounds caused by losses.

My Good Shepherd, I need Your anointing oil. I bow in submission, waiting for the touch of Your loving care.

Restore my hope.
   Renovate my emotions.
      Repair my attitude.
         Rejuvenate my purpose in life.
           Regenerate my wounds into wholeness.
              Rehabilitate my broken heart.

Thank You, Good Shepherd, for the restorative
power of Your oil of mercy.

> NEVERTHELESS WE MADE OUR PRAYER UNTO OUR GOD,
> and set a watch.
> *Nehemiah 4:9*

What gave Nehemiah and the people courage to face their enemies? Being prepared by being prayerful and watchful. What gives us courage in the face of death? Being prepared by being prayerful and watchful!

In many ways, we can never be prepared for death when it comes—whether slow and sure, or fast, or totally unexpected. Yet, as Christians, our mind-set should be, "Jesus may come today. Am I ready? I might die today! Lord, help me to prepare."

The loss of someone we love puts our bodies and minds into shock. Bringing to mind the sovereignty and goodness of God helps prepare our hearts to submit to His divine plan for us. Submission brings courage!

If we have been faithful in our commitment to Christ, we can readily trust Him during this most difficult time. Faithfulness gives courage to press on, and trust helps us to relax in the unchanging character of God.

Prayer without watching is hypocrisy;
and watching without prayer is presumption.

> FOR IN THE WILDERNESS SHALL WATERS BREAK OUT,
> and streams in the desert.
> *Isaiah 35:6*

Oh, Lord, I am in a desert and I need Your refreshing water! My tears do not refresh; they simply leave hollow depressions in the desert sand.

I need a stream of Your wise understanding
   in this desert of questions.
      I need a stream of Your soothing comfort
         in this desert of sorrow.
            I ask for a stream of Your kind blessing
               in this desert of hurt.
I yearn for a stream of acceptance
   in this seemingly illogical turn in my life.
      I must have a stream of undying hope if I am
         to make it through this desert of darkness.
      Broken dreams,
         Forfeited plans
            Drift away in these desert sands.

Please send some streams in the desert, Lord! Thank You.

> **WHEN JESUS SAW HIM LIE,** and knew that he had been now a long time in that case, he saith unto him, Wilt thou be made whole?
> *John 5:6*

Grief makes you sick! Sometimes you feel like a basket case. For many, it takes a long time to heal. It is okay to take time to heal, but there may come a time when you need to ask yourself, "Do I really want to get well? Do I really want to get over this grief?"

You wonder, "What will my family expect of me? What will be expected of me from the church? Can I again be a healthy, responsible member of society?"

For 46 years, I was the wife of a carpenter, the father of our children. For 43 years, I was the mother of our oldest daughter. Now—who am I? I am no longer a wife. I am still a mother, but I no longer have my eldest daughter.

It is difficult to re-identify myself. Sometimes I don't even want to. It's easier just to be a nobody. Odd as it may seem, there is something comforting about the pain of sorrow! You know others don't expect anything from you.

Even in the shattering experience of grief, there comes a time when you dissolve into a "comfort zone"—be it ever so painful. You feel "good" being cut off from the rush of "normal" living.

Beware, too much time spent in that painful "comfort zone" will render you useless and hopelessly selfish.

Yes, Lord, I want to be made whole
even though there is a hole in my heart.

> THE SEA IS HIS, AND HE MADE IT:
> and his hands formed the dry land.
> *Psalm 95:5*

(Since the sea is His, surely He will guide me across it.)

Jesus Savior, pilot me
Over grief's tempestuous sea.
Nameless thoughts and questions roll,
Chaos reigns and shocks my soul
Loneliness, and dread, and fears
Chart my path with sighs and tears.

Jesus Savior, pilot me
On this dark and stormy sea.
Sorrow's waves toss me about,
But I know You'll bring me out—
To that higher, safer ground
There by God's grace I'll be found.
—Faythelma Bechtel

*I cannot expect God's direction and protection
if I do not completely turn the helm over to Him.*

> **WHERE NO COUNSEL IS, THE PEOPLE FALL:**
> but in the multitude of counsellors there is safety.
> *Proverbs 11:14*

It has been said that a person in grief cannot think straight, and I think all who have been or still are in the valley of grief would agree. It is unwise to make any major decisions while the waves of anguish and sorrow are still dashing around your heart.

When you suddenly find yourself alone with a mountain of decisions to make, seek for some trusted counselors.

Finding a counselor or counselors to help you with decisions is not an easy task. Often pride keeps you in a private corner, or the fear of exposing yourself prevents you from asking for help. When you ask for someone to help you make decisions, you feel very vulnerable. You will feel vulnerable all during this valley experience—maybe that's one of God's ways of getting you to lean on Him (and others?).

Some decisions need a multitude of counselors, but decisions regarding your personal business usually require less. Knowing you have someone who will wisely direct your thinking will lift a great burden from your already-overloaded mind.

I thank God for the wisdom and help of my children.

> **WHY STANDEST THOU AFAR OFF, O LORD?**
> why hidest thou thyself in times of trouble?
> *Psalm 10:1*

"Why suffering?" is the age-old question of all mankind. In the book of Job, the suffering of Job was brought into the court of human reasoning. Job's wife first presents her reasoning. "Why do you keep your integrity, Job! Curse God and die. All this suffering is His fault!" (Let us remember, Job's wife was the distraught mother who had lost all her children, all the family wealth, and now it appeared she would lose her husband also.)

Blaming God is man's first and most common reaction to trials and loss. I wonder if the thanks God receives for all His blessings ever surpasses all the blaming and cursing He receives for man's trials and pain!

At that point in time, Job was too numb with grief and shock to blame God. He was anesthetized in the calm just before the storm. His questions and doubts and accusations would come later.

Oh, yes, how I remember that numb calm just before the storm. But thank God, He can calm the storm again and again.

Forgive me, Lord, when I blamed You for not answering the prayers (mine and the prayers of many others) that I thought should heal our daughter.

To live is to suffer; to survive is to find meaning in suffering.
—Victor Frankl, 1905-1997

> If I justify myself, mine own mouth shall condemn me.
> *Job 9:20*
> IF MY STEP HATH TURNED OUT OF THE WAY,
> and mine heart walked after mine eyes.
> *Job 31:7*

Job's visitors sat in shocked quietness before their afflicted friend. It was a custom for the visitors not to speak until the bereaved intimated a desire to be comforted.

After days of silence, Job was ready to boil over. Feelings buried alive will erupt! He began with the "why" questions. "Why died I not . . . why did I not give up . . . why did the . . . why is light . . . why hast Thou?"

Throughout Job's various speeches, we hear him questioning God and examining himself. If I had done this, or if I had done that, then God would have reason to punish me. *Cause me to understand where I have erred* (Job 6:24).

Yes, I've been in Job's place of self-examination and condemnation. The court of human reasoning can become very accusing and questioning. *This happened because....* Am I learning what God is trying to teach me?

A bit of self-examination is right and good. It alerts me to God's polishing process in my life. But let me never be like Job's friends and pass judgment on another.

I also must be aware that too much self-examination promotes guilt and discouragement and reduces faith and confidence in God's plan for my life.

> *Still as of old, men by themselves are priced—*
> *for thirty pieces Judas sold himself, not Christ.*
> —Cholmondeley

> STILL HE HOLDETH FAST HIS INTEGRITY,
> although thou movedst
> me against him, to destroy him without cause.
> *Job 2:3*

The battle was on—that forever earthly battle between good and evil—and Job was right in the middle of the crossfire. Why didn't God clue Job in on the reason for his tremendous suffering? Why didn't God tell Job he was a man of integrity? Why didn't God inform Job that this test by fire would prove to Satan that Job served God because he loved Him—not just because God blessed him?

We think that behind-the-scene knowledge would have saved Job a lot of anguish of mind, but would it have? Obviously, if such knowledge would have helped Job, God would have disclosed it to him. If Job had known how special he was to God, he would have had a whole set of different feelings to deal with.

The theology of Job's day was "the good prosper, and the bad suffer"—much like the health, wealth, prosperity thinking of our day. This bad theology of Job and his friends left Job bewildered and distressed in the court of human reasoning. God simply wasn't treating him as he thought he should be treated.

Along with Job, I often need to call out to God,
"Forgive these wild and wandering cries, forgive them when they
fail in truth, and in Your wisdom make me wise!"

> **SHALL HE THAT CONTENDETH WITH THE ALMIGHTY** instruct him? he that reproveth God, let him answer it.
> *Job 40:2*

Job had asked for an audience with God time and again. His trial in the court of human reasoning was disappointing to say the least. Suddenly, God granted Job's request, and Job found himself encircled with dozens of questions from Jehovah. The swirling questions helped refocus Job's mind. He saw the greatness of God in all creation.

Job saw the brevity of man's life as compared with God's existence: *Where were you when I made the earth?* (Job 38:4, free translation).

Job sensed the ignorance of man's reasoning compared with the omniscience of God: *declare if thou hast understanding* (Job 38:4).

In those questions, Job understood the helplessness of man compared with the omnipotence of God. *Canst thou lift up thy voice to the clouds* [and get rain]? (Job 38:34).

The voice of the Lord brought Job to the dust. *He cannot answer him* [God] *one of a thousand* (Job 9:3). Job had been contending with the Almighty, but now he has no questions and no answers.

God offered Job no answer as to why he was suffering. Humbly, Job took his place before the almighty God and realized he did not need to know the reason for his suffering. All he needed to do was to bow reverently in submissive trust to the plan and hand of God.

> *Suffering is the surest means of making us truthful to ourselves.*
> —Sismondi

> AND YE HAVE FORGOTTEN THE EXHORTATION which speaketh unto you as unto children, My son, despise not thou the chastening of the Lord, nor faint when thou art rebuked of him.
> *Hebrews 12:5*

The tears we shed are not in vain;
Nor worthless is the heavy strife;
If, like the buried seed of grain,
They rise to renovated life.

It is through tears our spirits grow;
'Tis in the tempest souls expand,
If it but teaches us to go
To him who holds it in his hand.

Oh, welcome, then, the stormy blast!
Oh, welcome, then, the ocean's roar!
Ye only drive more sure and fast
Our trembling bark to heaven's bright shore.
—Thomas Cogswell Upham, 1799-1872

We need to suffer that we may learn to pity.
—Landon

> THERE REMAINETH THEREFORE A REST to the people of God. For he that is entered into his rest, he also hath ceased from his own works, as God did from his.
> *Hebrews 4:9, 10*

The Heart at Rest

O blessed life—heart, mind, and soul
When all without tumultuous seems—
From self-born aims and wishes free,
That trusts a higher will, and deems
In all at one with Deity,
The higher will, not mine, the best.
And loyal to the Lord's control.
—William T. Matson, 1866 (adapted)

The grieving heart often finds it difficult to come to that place of rest. The mind is flooded with the disquietude and distress of many sudden changes. The mind is busy trying to figure out what to do about this, and how to take care of that. When we are so busy with our own work, God is unable to work.

What is it that gives us rest? Hope! And what is the source of our hope? Christ's resurrection! Jesus accomplished His Father's will by dying and rising again; and His resurrection is the foundation of our hope.

Hope is not a choice, but it is the result of making a decision to believe that God is in control of what is happening in my life. The more faith, the more hope. The more hope, the more rest!

Dying to my self-will is the great test—the test that develops faith—then hope and rest are added.

> THEY THAT TRUST IN THE LORD
> shall be as mount Zion, which cannot be removed,
> but abideth for ever.
> *Psalm 125:1*

Grief leaves you feeling very vulnerable and insecure. Before your grief experience, you knew that sustained stability never comes from circumstances. Circumstances are as changeable as the wind and weather.

One moment your circumstances may find you happy, healthy, and honored; and the next moment you may find yourself devastated and desperately depressed. If your anchor and hope are not in the unchangeable Christ, the billows of sorrow may take you under and keep you there!

Stability does not come from knowledge and education, as some would have you believe. Often the most educated person is the most unstable. His education has filled his mind with questions and doubts about everything and God in particular.

Grief, all kinds of losses, and severe trials come with their own unique questions. Do not make things worse by trying to make educated assumptions as to why things have happened to you as they have.

Stability does not accompany an educated mind that tries to figure everything out. The most stable things in life must be accepted by faith: God's love, Jesus' work on Calvary, the comfort of the Holy Spirit.

Sustained stability only comes from a firm faith and submission to the sovereign decisions of a loving God.

> GOD IS TOO GREAT TO BE WITHSTOOD,
> too just to do wrong,
> too good to delight in anyone's misery. We ought, therefore,
> quietly submit to his dispensations as the very best for us.
> —*Author Unknown*

The above quote must have been from a student who graduated from God's University of Adversity. Surely, we other Adversity students have experienced the same "tough" teaching, though it may have taken us a long while to realize God's best for us during the darkest hours.

» How often have I questioned or blamed God?

» How often have I felt God has forsaken me?

» How often have I stopped and asked myself, "Am I learning what God is trying to teach me?

» Am I submitted or resigned?

» Am I rebellious or surrendered to God's will?

» Have I accepted the changes in my life, or am I still fighting or denying them?

Commit yourself to the self-test of resignation or submission.

Lend Me Your Hope

Lend me your hope for awhile, I seem to have mislaid mine.
Lost and hopeless feelings accompany me daily,
Pain and confusion are my companions.
I know not where to turn.
Looking ahead to future times
Does not bring forth images of renewed hope.
I see troubled times, pain-filled days, and more tragedy.

Lend me your hope for awhile, I seem to have mislaid mine.
Hold my hand and hug me; listen to my ramblings;
Recovery seems so far distant.
The road to healing seems like a long, lonely one.

Lend me your hope for awhile, I seem to have mislaid mine.
Stand by me; offer your presence, your heart and your love.
Acknowledge my pain.
It is so real and ever present.
I am overwhelmed with sad and conflicting thoughts.

Lend me your hope for awhile,
A time will come when I will heal,
And I will share my renewal, hope, and love with others.
—Eloise Cole, used by permission

Hope is sort of like a spare tire. When nothing else motivates you to move ahead with life, hope will start your life back in motion—even if it is a temporarily borrowed hope.

The God of all hope offers you Himself.

> EVEN THE DARKEST,
> most lonely hour has only sixty minutes.

Yes, but sixty minutes hour after hour add up! Feeling alone can become overwhelming! What causes the feelings of aloneness? Loss of normalcy is one cause.

A widow feels alone even with children still at home, because she is a widow. A disabled mother and wife feels alone because she can no longer do what she used to do. A mother feels alone because she lost her first child. A once-happy wife feels alone because of her husband's unfaithfulness.

What makes your life abnormal? Maybe you can no longer go to work. Maybe you lost your mobility, lost your husband, lost your child, have a special needs child, or can't have a child—or one of a thousand other abnormal situations. No one among your friends is going through what you are right now. You are alone!

Many major losses result in a loss of your personal identity. Loss of identity causes you to feel alone. A widow or widower no longer has a companion to identify with. A rejected spouse finds it difficult to re-identify.

Sorrow and stress drag the mind out of normal thinking patterns, and you find yourself disoriented and confused. Your sense of direction is all tangled up in pain. Your initiative to move forward seems to have been buried with your loved one, or whatever loss you may have experienced.

I must remember I am never truly alone. I feel very alone at times because of the losses of normalcy, personal identity, and direction for my life. But, when I turn to God in my aloneness, I sense His presence and I can identify myself as His child.

This is possible only by God's grace, in God's time, with my acquiescence to His will.

> SEE WHAT *[an incredible]* QUALITY OF LOVE THE FATHER has given *(shown, bestowed on)* us, that we should *[be permitted to]* be named and called and counted the children of God!
> 1 John 3:1, Amplified

Does having valid reasons for that lonely feeling take it away? Indeed not! But reasons do add some saneness to our abnormal situations. It does help to realize that major losses put you into a different category than "normal," and in that "abnormal" category you are NOT alone!

So much depends on how I accept the place God has called me to.

I was born a daughter, later I became a sister, then I became a student, and then I chose to become a child of God.

God called me to be a teacher, then a wife, then a mother, then a mother-in-law.

Then God called me to be a grandmother and an aunt.

Now God has called me to be a widow!

Each of my callings before this last one, I accepted as good, as a gift from the hand of God. I endeavored to fill the place He gave me with the best of my feeble ability, and I accepted the challenges of each calling. Can I do less with my calling to be a widow?

As God's child, I have His assistance in
learning how to be a widow for God's glory!

> WHEREFORE SEEING WE ALSO ARE COMPASSED ABOUT WITH so great a cloud of witnesses, let us lay aside every weight, and the sin which doth so easily beset us, and let us run with patience the race that is set before us.
> *Hebrews 12:1*

When you're feeling so very alone, read this verse over and over. You are on earth's stage being watched and cheered on, not only by other earthly pilgrims, but also by those on the grandstands in heaven—earthly terms for a heavenly scene.

A great "cloud of witnesses" is interested in what is going on in your life and mine. Surely we don't want to be a disappointment to them, and most of all, we don't want Jesus to be disappointed in us.

When you become a widow or a loser through some other tragedy, you often feel as though you are on display. Some of the most common things said or written to me have been, "I hope you're finding God's grace sufficient. Are you finding God's grace sufficient? God's grace is sufficient!"

Yes, we are surrounded with observers! How are we displaying the grace of God to others? Sometimes we simply do not feel God's grace. There have been times when I have needed to lay aside the weight of my loss and embrace God's grace. Thinking of the joy and complete healing of my loved ones in heaven, possibly observing me gives me, that courage to keep pressing on.

Joys are our wings; sorrows our spurs.
—Richter

Joy is more divine than sorrow, for joy is bread and sorrow is medicine.
—Henry Ward Beecher, 1813-1887

> W̲ITH MEN THIS IS IMPOSSIBLE;
> but with God all things are possible.
> *Matthew 19:26*

But God

I know not, but God knows
The "whys" of all my pain;
Even my future days
To Him are clear and plain.

I cannot, but God can
Carry my burden sore;
When strength and courage fail,
He grants me more and more.

I see not, but God sees;
For He provides the light
And gently leads me on
In peace—His presence bright.

I claim not, but God claims
My future in His care—
Makes all things possible,
Even calls me His heir.
—Faythelma Bechtel

> HE ANSWERED AND SAID, LO, I SEE FOUR MEN LOOSE, walking in the midst of the fire, and they have no hurt; and the form of the fourth is like the Son of God.
> *Daniel 3:25*

Have you felt the presence of God in the dark valley? There likely were times when you did not feel like He was there, but later you realized He had to have been there with you or you would not have made it through those darkest hours.

He was with the three Hebrew men in the fiery furnace. I doubt that the men expected to meet Christ in the fire, but they believed God would deliver them one way or the other—through the miracle of life or the transformation by death.

God is always present to deliver us and to bring glory to His name. His presence secures our deliverance. By His sympathy, He increases our courage. By His strength in our affliction, He increases our strength.

God does not always prevent us from falling into distressful situations, but He is always there to offer support, comfort, and direction in our troubled times. Jesus did not turn down the heat in the furnace, but He did accompany them through the fire.

It is in the dark valley and fiery furnace that the character of God is unveiled, and the wonders of eternity come near.

> **WHAT MANNER OF COMMUNICATIONS ARE THESE** that ye are having one to another, as ye walk, and are sad?... What things?
> *Luke 24:17, 19*

Jesus, the all-knowing one obviously knew "what things" had transpired and what had so greatly saddened these two on the way to Emmaus. So, why did He ask what their conversation was about?

Jesus, the one who hurts when we hurt, was inviting them to share their painful hearts with Him. He wanted them to express their innermost feelings, doubts, discouragements, and fears concerning the recent tragedy in their lives.

Remember: emotions suppressed = distress;

emotions expressed = rest.

This is especially true when the emotions are shared with the Man of Sorrows, the only One who can truly understand all about our pain.

After the sorrowing men shared their frustrations and wonderings, Jesus "opened their understanding." He assured them that this tragedy had happened for their own good.

Thank You, Lord, for inviting me to share my pain. Please open my understanding which throbbing grief has greatly clouded. Sift out the bitter reasoning from the better and more beneficial thinking.

As I cast my doubts and fears on You, help me to find rest.

> I AM THE RESURRECTION, AND THE LIFE.
> *John 11:25*

It is not death to die;
To leave this weary road,
And 'midst the brotherhood on high
To be at home with God.

It is not death to fling
Aside this sinful dust,
And rise, on strong exulting wing,
To live among the just.

Jesus, Thou Prince of Life!
Thy chosen cannot die;
Like Thee, they conquer in the strife,
To reign with Thee on high.
—Henri Malan, 1787-1864

When we look at death through the eyes of Jesus, it is painless.

Death is life's greatest victory.
Death is life's greatest paradox.
To live is to die, and to die is to live.

> And he said unto me, MY GRACE IS SUFFICIENT FOR THEE: for my strength is made perfect in weakness. Most gladly therefore will I rather glory in my infirmities, that the power of Christ may rest upon me.
>
> *2 Corinthians 12:9*

S - strength
U - unending
F - fulfilling
F - fellowship
I - infinite
C - comfort
I - intense
E - expectations
N - never-ending
T - trust

When I am weak, then Jesus shows Himself strong. Someone has said "grace humbles a man without degrading him and exalts him without inflating him."

Grace is the ability to do things God's way, to see through His eyes, to feel with His heart, and to understand with His wisdom. Grace is simply God working in and through my frail humanity—a miracle indeed.

"I know it is only Your grace that will bring me through this dark, painful valley. You brought me through it once, and you can do it again and again!"

Happiness keeps you sweet.
Trials keep you strong.
Sorrows keep you human.
Failures keep you humble.
Success keeps you growing.
But only God keeps you going.
  —Author Unknown

> **I KNOW THAT, WHATSOEVER GOD DOETH,**
> it shall be for ever: nothing can be put to it, nor any thing taken from it: and God doeth it, that men should fear before him.
> *Ecclesiastes 3:14*

How often have you wished to hurry God, slow Him, or even to stop Him? Have you ever wished to add to His plans or take away from His plans? Have you observed the perfection in all of God's creation, and then thought of things you would have done differently?

One man was so bold to say, if only the Almighty had called him into counsel at the making of the universe, he could have given the Almighty some valuable hints.

We very quickly recognize this as a mouthful of foolish arrogance. Yet, how often have you and I wished God had managed our lives differently?

"Help us, Lord, to acknowledge Your work as perfect, unalterable, purposeful, and eternal. Even in our losses, Lord, You have a purpose—'that men should fear before him.'"

May we ever remember it is the perfect, unalterable, purposeful, and eternal work of God that is the foundation for our hope and the reason for our trust.

> TO EVERY THING THERE IS A SEASON,
> and a time to every purpose under the heaven.
> *Ecclesiastes 3:1*

In every grief there is a season.
A time to be numb, and a time to be hypersensitive;
A time to weep, and a time to stop weeping;
A time of blackness, and a time of light;
A time to block memories, and a time to inhale memories;
A time to shut yourself away, and a time to emerge;
A time to deny your feelings, and a time to face your feelings;
A time of questions, and a time of accepting no answers;
A time of hopelessness, and a time of faith and hope;
A time of feeling forsaken, and a time of feeling secure in God;
A time of lost identity, and a time to identify with Jesus;
A time of confusion, and a time of comfort;
A time to care for yourself, and a time to reach out to others;
A time of rebellion, and a time of submission;
A time of heartbrokenness, and a time of healing.
—Faythelma Bechtel

You may have more, less, or different times than these, but to every time there is a purpose. It is not in man to understand all the purposes or times of God, for that would equate him with God.

Help me, Lord, to find You in every purpose.

> NOW FAITH IS THE SUBSTANCE OF THINGS HOPED FOR,
> the evidence of things not seen.
> *Hebrews 11:1*

The impossibility of my valley experience producing something good sometimes hangs over me like a dark cloud. For tears to be transformed into triumph, for heartbreak to become heart healing, for sorrows to turn into joy, all are impossible for me to imagine, much less to accomplish. *But with God all things are possible!* (Matthew 19:26). Increase my faith, Lord!

Faith is evidence of things not seen? Only in God's world by faith is it possible to see the invisible! Hebrews 11:27 says, *Moses endured as seeing him who is invisible.* Moses' faith in his invisible leader kept him moving forward. His faith in the invisible gave him hope and courage to go on.

For our light affliction, which is but for a moment, worketh for us a far more exceeding and eternal weight of glory; While we look not at the things which are seen, but at the things which are not seen: for the things which are seen are temporal; but the things which are not seen are eternal (2 Corinthians 4:17, 18).

In these verses we again see a paradox of the God of impossibilities. When faith is most difficult, that is when it is most necessary.

Help me, Lord, to keep my mind and eyes focused on
the invisible glories of the eternal.

> I HAVE HEARD THEE IN A TIME ACCEPTED,
> and in the day of salvation have I succored thee: behold, now is
> the accepted time; behold, now is the day of salvation.
> *2 Corinthians 6:2*

Have you ever felt, *I can't make it another week? Another month? Another year?* You think about the recent past and feel anguish. You think about the future and feel trepidation. You are living in the anxiety of now.

How far ahead does God expect us to "make it"? Does God encourage us to live in the past? Does He want us to try to forecast the future? No!

We have a God of the NOW. Today is the time we need to deal with. And we need deal with it only a moment at a time.

It is all right to imagine God saying, "Now is the day of salvation—from your fears, your questions, your pain, your depression, your anger, or whatever you face in the valley. I will help you now—today! My grace is sufficient for you now—today! I offer you comfort and compassion and forgiveness now—today!"

T—traveling
   O — onward
      D — daily
         A — assisted by
            Y — You, Lord!

Do not look back on happiness or sadness,
or dream about the future or dread it.
You are only sure of today; do not cheat yourself out of it.

God's ABC's for Losers

When the valley seems a desert, may you Abound in His Abundance.
In famine times, may you awake to the Bliss of His Blessings.
In lonely hours, may you wrap yourself in the Comfort of His Compassion.
On depressing days, may you Delight in your Dependence on Him.
When discouraged, may you be Encircled with His Encouragement.
When fearful, may you trust His Fathomless Faithfulness.
When tempted to complain, think on the Generosity of His Goodness.
When apathetic, may you experience the Hyssop of His Healing.
When your mind is full of questions, remember the Infallibility of His Integrity.
When you feel ridiculed, contemplate the Jurisdiction of His Justice.
When you feel abandoned, consider the Kindness of His Keeping.
During times of darkness, may you walk in the Light of His Love.
When you feel like a total failure, may you sense the Magnitude of His Mercy.
When you walk in self-confidence, may you remember the Necessity of His Nearness.
When you feel helpless, may you experience the Overflowing of His Omnipotence.
When in turmoil, commence your Quest for His Quietness.
When your life seems fruitless, Rely on His Righteousness.
In your weakness, may you rest in the Strength of His Stability.
When struggling with doubt, Treasure His Truths.
When you feel unsettled, think on the Uniqueness of His Unchangeableness.
In times of defeat, grasp the promise of the Verity of His Victory.
When life is full of uncertainties, seek the Wealth of His Wisdom.
Never feel vulnerable under His X-ray of love.
There is no risk in Yielding to His Yoke.
May you accept His enrichment in your life with Zest and Zeal.

—Faythelma Bechtel

> THE FEAR OF THE LORD IS THE BEGINNING OF KNOWLEDGE:
> but fools despise wisdom and instruction.
> *Proverbs 1:7*

In this verse, I hear God offering me knowledge that will instruct me on how to ride the waves and dock safely during the storm.

The fear of God is like a lighthouse on the rocky cliffs surrounded by an angry, stormy sea, sending out bright beams that help guide the weary sailor to the shore safely.

- » The fear of the Lord beams rays of reverence for my God's omnipotence.
- » The fear of the Lord beams rays of obedience to His commands.
- » The fear of the Lord beams rays of understanding His compassion.
- » The fear of the Lord beams rays of acceptance of His perfect will.

Thank You, Lord, for these beams of knowledge that are directing my voyage.

I will remember,
"Smooth seas never make skillful mariners."

> SO WHEN THIS CORRUPTIBLE SHALL HAVE PUT ON incorruption, and this mortal shall have put on immortality, then shall be brought to pass the saying that is written, Death is swallowed up in victory. O death, where is thy sting? O grave, where is thy victory?
> *1 Corinthians 15:54, 55*

Oh, the terrible sting of death
If there was no hope beyond the grave;
But praise be to Christ for His
Death and Resurrection pow'r to save—

By His offer of great grace
That kills the sting of finality;
Giving us eternal hope
And bestowing immortality!

But I would not have you to be ignorant, brethren, concerning them which are asleep, that ye sorrow not, even as others which have no hope. For if we believe that Jesus died and rose again, even so them also which sleep in Jesus will God bring with him.
1 Thessalonians 4:13, 14

O God, deliver me from the ignorance of sorrowing as those who have no hope! My hope is in You!

> I WILL GO IN THE STRENGTH OF THE LORD GOD:
> I will make mention of thy righteousness, even of thine only. O God, thou hast taught me from my youth: and hitherto have I declared thy wondrous works. Now also when I am old and grayheaded, O God, forsake me not; until I have shewed thy strength unto this generation, and thy power to every one that is to come.
> *Psalm 71:16-18*

What gives me my go-power? Candy? Eating? Talking about me and my pain? Sleeping to escape my hurts?

What gives me my go-power? Praying? Talking about God's righteousness? Talking about God's wonderful works? Remembering God's unfailing help in the past? Waiting on God and trusting in His promises?

The verse says, "I will go!" First, I must have *resolve* to go—to plunge on through the valley and on to the mountaintop. Then I must have *confidence* in God's strength and never trust my own. Then I must have a *purpose*. I purpose to declare His righteousness. I purpose to tell of His wondrous works. I purpose to show His strength and power to the next generation.

What a challenge, Lord! May the next generation not know me as a loser, but a winner. May they see my tears turned to pearls. May they not remember my heartbreak; but may they see my heart refurbished with compassion for others.

O Lord, I will go in Thy strength!

> CIRCUMSTANCES MAKE A PERSON NEITHER STRONG NOR WEAK;
> they only show which he is.
> Do you look through your difficulties to God—
> or, through God at your difficulties?"

Lord, all that I should have said,
  All that I should not have said,
    All that I should have done,
      All that I should not have done,
I bring all to You, Lord, and ask for forgiveness.
  All that I couldn't be,
    All that I wouldn't be,
      I now confess to You.
        All promises that were broken,
          All hopes that were dashed,
            All dreams that never happened,
              All plans that were slashed,
I place at Your feet and ask to be made complete.
  All of my failures, all of my fears,
    All of my joys, all of my tears,
      All of my pride, all of my doubt,
        All of my thanks, all of my pout,
I place in Your hand
  And by Your side I take my stand
    As a forgiven and loved child of Yours.
        —Faythelma Bechtel

As a Mother Comforteth

Where'er dumb sorrow hopeless lies,
And, lost to anguish, tear-dimmed eyes
Shun the "good morning" on the skies,

Speak comfort, Lord, with every breath,
Yea, as a mother comforteth,
Bind up and heal the wounds of death,

Then pity, till Thy mourners see,
In Thine own pierced hand, the key
To every life-long mystery;

Until the dawn break more and more,
When, the last strife and heartbreak o'er,
The last storm spent upon the shore,

Death shall be vanquished at Thy will,
And good, eternal, follow ill
At Thine omnipotent "Be Still"!

—Mary Rowles Jarvis, 1880

Do not seek death. Death will find you.
But seek the road which makes death a fulfillment.
—Hammarskjöld

> BUT I AM POOR AND NEEDY;
> yet the Lord thinketh upon me:
> thou art my help and my deliverer;
> make no tarrying, O my God.
> *Psalm 40:17*

In any circumstance, the word *separation* brings thoughts of tenderness, loneliness, something missing, and pain. After a death, separation is uppermost in our minds and is more than a gnawing, unidentified, indefinite pain. It is a hole in our hearts with pain that is indescribable.

We are left on earth, and our loved one is in heaven. With that separation comes feelings of aloneness, instability, and abandonment, without sense of purpose or direction.

Who did God have in mind when He inspired Romans 8:37-39? *Nay, in all these things we are more than conquerors through him that loved us. For I am persuaded, that neither death, nor life, nor angels, nor principalities, nor powers, nor things present, nor things to come, nor, height, nor depth, nor any other creature, shall be able to separate us from the love of God, which is in Christ Jesus our Lord.*

God knew we were going to face difficult separations on this earth. He knew we were going to need some connections for stability, security, and assurance.

God is that immovable, trustworthy connection
who is always only a prayer away.

The Changing Scenes of Life

The changing scenes of life
Sometimes create a mountaintop,
The views of beauty are so great
We wish that they would never stop.

The changing scenes of life
Sometimes crush us in a corner,
Sometimes they stunt and cripple us
Or abandon us—the mourner.

The changing scenes of life
Produce aches and growing pains,
But our God of all compassion
Sends what is best of suns or rains.

The changing scenes of life
Oft take us to the valley low,
There we are challenged to survive;
There we submit, commit, and grow!
—Faythelma Bechtel

Life brings change! Some changes are brutally painful, and others are delightfully beautiful. All changes, positive or negative, bring stress into your life—positive or negative stress!

How I handle stress is an expression of how I utilize God's grace or decline it in my life (excluding physical deficiencies).

Hold my hand, Lord, through all life's changes!

> And Elisha prayed, and said,
> LORD, I PRAY THEE, OPEN HIS EYES,
> that he may see. And the LORD opened the eyes of the young
> man; and he saw: and, behold, the mountain was full
> of horses and chariots of fire round about Elisha.
> *2 Kings 6:17*

Lord, I've been traveling through this dark valley for a long time. The path has been steep and winding. Sometimes I feel like I'm back where I started, but I'm not, for I am on higher ground.

When life seemed hopeless, You lifted up my eyes to my help on the mountain. When life was filled with dismay and fear, You opened my eyes to my help on the mountain. When I felt like giving up, I saw Your horses and chariots waiting to help me.

You have kept me safe through the dark valley and carried me to a higher plane. This difficult journey has made me stronger. Thank You, Lord!

And Lord, continue to bless all those weary travelers that have gone before me and all those who are following after me. Open their eyes to Your help that comes from the mountain.

The storms of life not only confirm
my strength, but also increase it.

> I WILL LIFT UP MINE EYES UNTO THE HILLS,
> from whence cometh my help.
> *Psalm 121:1*

What Is in a Name?

As you carefully read over each name, close your eyes and picture how Christ has been that to you during your grief experience.

He is my rock, my fortress, my judge,
   my king, my shield, my companion,
      my deliverer, my refuge, my Savior
         my shepherd, my hope, my keeper.
He is the light, the life, the truth, and the way.
   He is grace, mercy, peace, wisdom, and love.
      He is the great *I am that I am.*
That means,
   He is ALL that is necessary for any occasion that will arise.
   He is the eternal, unchangeable God! When you add the immutable character of God to this (incomplete) list of names, that increases our security, confidence, and assurance that He has been, is, and always will be there for us.

It is impossible to have a too exalted
concept of our great God.

> HAST NOT THOU MADE AN HEDGE ABOUT HIM,
> and about his house, and about all that he hath on every side?
> *Job 1:10*

Satan could not touch Job because of the hedge of protection that surrounded him. Only with God's permission did Satan get inside that hedge and leave Job's life in a state of chaos and pain. Because of that hedge of protection, Satan was unsuccessful at getting Job to curse God.

I like to think of that hedge as God's unchangeable character. This hedge does not protect me from the painful trials of life. This hedge does not secure for me all the answers to my questions. This hedge does not offer instant relief or quick fixes for my anguish.

It is a hedge of His compassion and wisdom; I can trust Him in the direction He is taking my life. It is a hedge of hope and peace; I can rest in the knowledge that He knows what is best for me. It is a hedge of love and justice; I can depend on Him not to give me what I deserve.

For I am the Lord, I change not; therefore ye sons of Jacob are not consumed (Malachi 3:6). Because God is constant and consistent, I will not be consumed by my grief or any other difficulty I face. I can count on Him always to act according to His name.

Nothing harmful can pass through the hedge of His
unchangeable character except with His divine approval.

> The Lord hath appeared of old unto me, saying,
> YEA, I HAVE LOVED THEE WITH AN EVERLASTING LOVE:
> therefore with lovingkindness have I drawn thee.
> *Jeremiah 31:3*

To love and to be loved
Is to hurt and to be hurt!
Shall I say "no" to love
And avoid the hurt?

No love in life equals
No life to live;
It is a solitary death without love—
The most agonizing hurt life can give.
—Faythelma Bechtel

Love is never lost.
If not reciprocated, it will flow back
and soften and purify
the heart of the giver.

> WHO CAN JUDGE OR MEASURE GOD'S ABILITY
> to do according to His capacity?

The book of Ephesians lets us take a peek into our heavenly bank account. I hope this blesses you as it did me! Let's not ignore our wealth, but take advantage of it and grow.

» I receive spiritual blessings *according as he hath chosen* (me) *before the foundation of the world"* (1:4). Can I fathom that?

» I am an adopted child of Christ *according to the good pleasure of his will* (1:5). I have a Father who cares!

» I have redemption and forgiveness *according to the riches of his grace* (1:7). That's incalculable!

» He keeps me walking in the light that illumines the mystery of His will (the mystery of being a widow?), *according to his good pleasure* (1:9). My pain is not His pleasure, but His purpose is His pleasure.

» I have an inheritance *according to the purpose of his own will* (1:11). Can I comprehend that?

» He raised me to a new life as He raised His Son from death *according to the working of his mighty power* (1:19).

This inconceivable power can overcome all obstacles—even death.

> WHO CAN JUDGE OR MEASURE GOD'S ABILITY
> to do according to His capacity?

» I am a partaker of His promise and a minister *according to the gift of the grace of God given unto me* (3:7). Thank You, God, for the gift of grace that oils the stiff testings of life. Make my life a story of Your grace.

» I have fellowship with God and experience His wisdom *according to the eternal purpose which he purposed in Christ* (3:11).

» He strengthens my inner spirit *according to the riches of his glory* (3:16). I cannot calculate the riches of His glory!

» He is able to do more than I can ask or think *according to the power that worketh in us* (3:20). My faith is the measuring stick here, Lord!

» He gives me grace *according to the measure of the gift of Christ* (4:7). That measure is as great as every need! Thank You, Lord.

» I am a part of the body of Christ *according to the effectual working of every part in love* (4:16). Truly, I am never alone when I belong to Your body. Make Your love shine through my life!

Thank You Lord, for my heavenly bank account!
Help me to live according to Your riches and to the praise of Your glory.

My journey through the valley of the shadow of death has taught me:

» I serve a God of divine power. *Now I know that You can do everything—whatsoever pleases You.*

» I serve a God of divine knowledge. *No thought can be withheld from You.*

» I serve a God of divine purpose. *The Lord turned the captivity of Job.* In His time, God will heal and restore and prove Satan defeated.

» I serve a compassionate God. My suffering has opened my eyes to my own smallness and weakness and opened my heart to God's greatness and strength. *I have heard You with my ear, but now my eye sees You.*

» I serve a forgiving God. He cuts down my pride to make room for His grace. *I repent in dust and ashes.*

» I serve a just God. He takes away only to enrich, never to ruin and destroy.

It is one thing for God to be supreme, and another thing for man to know that He is. For our own good, for our guidance and assurance, and for God's glory, it is essential that we acknowledge the greatness of God. The valley can be the most excellent and the most painful school for learning that truth.

If we spend sixteen hours a day dealing with tangible things and only five minutes a day dealing with God, is it any wonder that tangible things are two hundred times more real to us than God?
—William Inge

> O THE DEPTH OF THE RICHES AND WISDOM
> and knowledge of God!
> How unfathomable *(inscrutable, unsearchable)* are His judgments—His decisions! And how untraceable *(mysterious, undiscoverable)* are His ways—His methods, His paths! For who has known the mind of *[the]* Lord and who has understood His thoughts, or who has *[ever]* been His counselor? Or who has first given God anything that he might be paid back or that he could claim a recompense? For from Him and through Him and to Him are all things. —For all things originate with Him and come from Him; all things live through Him, and all things center in and tend to consummate and to end in Him.
> To Him be glory forever! Amen—so be it.
> *Romans 11:33-36, Amplified*

Oh, Lord, when I consider who You are, I realize I must surrender all my questions to You. "Why have You taken my loved ones? What do You want from my life? Where do I go from here? How do I go on living? When will this painful journey be finished?"

Every why, what, where, how, and when question I give to You, Lord! Take them, and tuck them away in Your great mind. Then take me in Your arms, and let me find sweet rest in trusting confidence that You are in charge!

I need not a rest from work, but a rest in work;
rest and confidence in God's work in and through my life.

> FEAR THOU NOT; FOR I AM WITH THEE: BE NOT DISMAYED;
> for I am thy God: I will strengthen thee; yea, I will help thee;
> yea, I will uphold thee with the right hand of my righteousness.
> *Isaiah 41:10*

Lord, be my peace when I am uptight.
  Be my strength when I am weak.
  Be my courage when I am fearful.
   Be my companion when I am lonely.
    Be my ear when all is silence.
     Be my judge when others criticize.
      Be my mediator when others misunderstand.
     Be my defense when I am put on trial.
     Be my confidence when I'm in confusion.
    Be my fullness when I am empty.
   In my aimlessness, be my All in All!

I stretch forth my hands unto thee...
Psalm 143:6

Fill up the emptiness of your heart
with love for God,
(for your family), and your neighbor.
—Edith Stein

> BUT MY GOD SHALL SUPPLY ALL YOUR NEED according to his riches in glory by Christ Jesus.
> *Philippians 4:19*

God has promised to supply all my needs—so I feel I need someone to call me and say they're thinking of me, but no one calls. I feel I need someone to stop by so I can share my deep feelings of loneliness and sadness instead of bouncing my words off the non-feeling walls, but no one stops by. I feel the need for a hug or just a light in the eyes of someone who cares, but there's no hug or light of caring. Has your mind ever traveled along these lines? Mine certainly has!

God, where are You? What about my needs? I am learning that God looks at my needs in very different ways than I do. As soon as I accept God's silence, I also see my need in a very different light. Later, I find I am very glad no one stopped by when I was at my lowest. I am thankful no one but God heard me crying my heart out. I am glad no one called and listened to my babble. Thank You, Lord, for saving me the embarrassment of exposure so many times when I thought I needed someone and found I only needed You!

Each time I tell God I have a need and His silence says "no," and I accept His *stand still and see my salvation,* I am blessed by fulfilling His will and squelching my own.

Help me to remember that Jesus found peace in accomplishing His Father's will by accepting the crown of thorns, not by seeking comfort from humans who couldn't possibly understand anyway.

The strength of a man consists in finding out the way God is going, and going in that way too.
—Beecher

> Someone has said,
> "LONELINESS IS HAVING WHAT YOU
> DON'T WANT, OR WANTING WHAT YOU DON'T HAVE."

I don't want to be alone, but I am. I want companionship, but I don't have it.

Is my loneliness as a scorching, hot, sand-blown, dry desert of nothingness? "All day I face the barren waste without a taste of water, cool, clear water!" Or is my loneliness as a shady, sweet-smelling forest through which I travel, finding delicious berries, delightful wildflowers, and lush green places to rest?

What makes the difference in how I find my loneliness affecting me? Does the difference not lie in my resignation or my acceptance of where God has put me? When I resign, I wither and die; when I accept, I find peace and growth.

Acceptance results from having a vision of hope for the future—not the immediate future, for that can be very frustrating and frightful. But I must have a vision of hope for my eternal future.

Rather than ask God, "Why have You left me in this lonely position?" it would be better to say, "During these lonely, quiet times, teach me, Lord, more about Yourself and Your will."

*Limitations are God's caution signs that
encourage us to lean on His shoulder.*

> HEAR ME SPEEDILY, O LORD: MY SPIRIT FAILETH: hide not thy face from me, lest I be like them that go down into the pit.
> *Psalm 143:7*

That sounds like the prayer of a desperate man. It's a man who is saying, "Lord, if You don't listen to me, You're going to lose me." Is that a threat? Is there blame in the tone? Is there self-pity in the heart?

When I fail to accept what God has allowed in my life, I tend to find fault with God and blame Him for my pain, my loneliness, and my trials.

Finding fault with God raises a wall between God and me. Then comes that empty, totally alone, and depressed feeling of being forsaken by God.

God is great enough and good enough that He could easily fix my lonely problem. He could remove my trials, but that evidently is not in His plan for my life.

Acceptance of His plan (and whatever that may entail) is the golden staircase that keeps me from the dumps of aloneness and self-pity and keeps me climbing toward my heavenly home.

Whenever I find my prayers becoming desperate and demanding, help me, Lord, to refocus on You.

> **LONELINESS IS A GIFT GIVEN TO ME WITHOUT REQUEST;**
> I sacrifice it back to Thee, Lord, for peace and rest.

Can I say, all that I am, all that I have, all that I do is Christ's? That is a living sacrifice. Loneliness and sorrow are now a large part of me. Can I sacrifice them back to God? It is impossible to give back to the Lord what we have not accepted.

Have you ever given a child a dollar to spend on someone just so the child could experience the joy of giving? How often does Christ give me something just so I can experience the joy of giving it back to Him?

Do I see my singleness as a handicap, a liability, or a freedom ticket—free to do whatever pleases me? Or do I see my singleness, my loneliness, as gifts from God. Do I accept them as gifts and offer them back on the altar of consecration?

What God can do with such meager sacrifices (sacrifices that no one else would recognize or understand) is not really my concern. My concern is that my acceptance of His gifts, and my sacrifice will bring Him glory in some miraculous way.

*Life's trials are God's educational
system to develop faith.*

Which Bible characters impress you most? Who teaches you the most about God's greatness and power? Whose life would you like to pattern your life after?

Job lost all his riches, his family, and his health. Satan put him on trial and waited for him to curse God and die.

God asked Abraham to leave family and homeland and go to a land foreign to him. Abraham waited dozens of years for God to fulfill His promise of a son, and then God asked him to sacrifice his son.

Moses spent forty years in Pharaoh's court—away from his family, 40 years alone in the desert tending sheep, 38 years of wilderness wanderings, leading a murmuring, blaming, rebellious bunch of people.

Naomi's husband and sons died in a heathen land. Naomi returned to her homeland with a grieved heart and a lone daughter-in-law.

Joseph was hated by his brothers and sold into slavery. He was accused of sin by his master's wife, put into prison, and forgotten. He went from the pit, to the prison, to the palace in God's time.

As a youth, Daniel was taken into captivity in a foreign land. He went from the palace to the lion's den.

All of these characters had one thing in common—they knew God in a real, living way. They faced trials and impossibilities in their lives. They experienced God's power and direction and grace in their anguish.

The characters I admire are those who went through hard things. The characters I want to be like are those whose faith was severely tried and survived.

Am I ready to be tried? Will my faith survive?

> AND I WILL GIVE YOU THE TREASURES OF DARKNESS and hidden riches of secret places, that you may know that it is I, the Lord, the God of Israel, Who calls you by your name.
> *Isaiah 45:3, Amplified*

If God had a special work and special treasures waiting for heathen King Cyrus, how much more might He have waiting for His dedicated children? What are His treasures and riches hidden in the deep shadows of the dark valley?

Treasuries were built without windows for greater security. Truly, treasures were hidden in darkness. It has been estimated that just one of the treasures that fell into the hands of Cyrus from Babylon was worth more than one hundred and twenty-six million in sterling silver.

Such earthly treasures mean little when you have lost someone you love. Yet, is there a lesson for us here in the gloomy darkness, in secret places?

We are told that the famous lace shops of Brussels spin their finest lace and most delicate patterns in darkened rooms. One very small window admits a shaft of light which falls directly on the pattern. There is only one spinner in the room. He sits where that shaft of light falls on the threads he is weaving.

This is how we get our choicest products—one weaver in the dark and only His pattern in the light.

When I can't see, when I can't understand, keep me weaving, Lord!

Weaving

Grieving is weaving a tapestry of pain.
The yarn is pure black. The design sternly plain.
The toil terribly tedious, no end is in sight.
Work begins in the morning,
Goes on through the night.
Each weaves alone on his personal shroud,
Locked in the closet—away from the crowd.
When finally completed, it's tucked safely away
To be occasionally pulled out on a dark, dreary day.
—Author Unknown

Grieving is weaving a lace fragile and fine;
On the darkest days the pattern's His—not mine.
The Master Weaver sheds a shaft of heavenly beam;
Blending some golden strands
In an astounding scheme.
Each grief and each trial adds marvelous shades.
While I weep—God smiles as great splendor cascades.
Only God changes ashes for beauty, gloom for light.
He alone weaves robes fit for eternity's delight.
—Faythelma Bechtel

> Nay, in all these things
> WE ARE MORE THAN CONQUERORS THROUGH HIM THAT LOVED US.
> *Romans 8:37*

Does God really expect you and me to be more than a conqueror when we're held in sorrow's vise? Does He expect us to be more than a conqueror when we've had major losses? How can a loser be a conqueror?

When an army is more than conqueror, the army takes spoils—foods, supplies, and possessions. That is just what this verse means—there are spoils to take.

The story is told of Dr. Moon, who when stricken with blindness said, "Lord, I accept this talent of blindness from Thee. Help me to use it for Thy glory, that at Thy coming, Thou mayest receive Thine own with usury."

God enabled him to invent the Moon Alphabet for the blind. Because of his work, thousands of blind people were able to read God's Word, and many were saved.

That's being more than a conqueror! Dr. Moon accepted what God sent him and offered it back to God with spoils.

Turn the storm clouds into a chariot, my child;
Ride as conqueror, the tempest dark and wild.
Up, on eagles' wings now soar above all thy pain,
Great the spoils for thy Master thou shalt ascertain.

> **MANY ARE THE AFFLICTIONS OF THE RIGHTEOUS,**
> but the LORD delivereth him out of them all.
> *Psalm 34:19*

Have you ever asked, "Why do bad things happen to good people?" That's a common question of mankind. We think we deserve better than what God is giving us. According to God's Word, we are not good people, and He treats us better than we deserve.

Too many people think that being a Christian brings immunity to trials and losses. God offers no benefit package, except the forgiveness of sins, His presence and power through rough times, and a home in heaven to those who remain faithful.

Though he were a Son, yet learned he obedience by the things which he suffered (Hebrews 5:8). Can you explain that one?

It has been said, "It is because we are so imperfectly righteous that we have to suffer many afflictions." We have looked at our grief from the angle of living in a sinful world. Now we must look at it from the painful angle of being a sinful people.

God does not generally take our loved ones because we are sinful. But God certainly uses such painful experiences to purge, purify, and instruct our imperfect thinking and behavior. When our afflictions have accomplished His appointed work, God begins His work of deliverance.

He delivers from anger, bitterness, hopelessness, fear—and oh, so many possible pitfalls in the dark valley. But, as I yield to His plan and submit my pain to His loving care, I begin to feel deliverance. It will be a heavenly day when He will deliver me out of them all.

Though he were a Son, yet learned he obedience by the things which he suffered.

—Hebrews 5:8

> **THOU HAST ENLARGED MY STEPS UNDER ME,**
> that my feet did not slip.
> *Psalm 18:36*

When I think of stepping stones, I think of how different they can be. I have some small, tippy stones through my flower garden that are not the safest to walk on—especially on a dark night. But some stones are large and comfortable to walk on. Large, comfortable stepping stones provide security.

The idea of enlarged steps through the dark valley appeals to me and comforts me. It also caused me to search Psalm 18 to find how God has enlarged my steps.

He has enlarged my steps by making His strength available. He provides deliverance from fears by His presence. He gives light through the darkness. His delight in me gives me courage. His mercy, gentleness, and salvation are my security. Because He himself is the firm Rock beneath each of my steps, I will not slip if I continually trust Him and allow Him to lead me through this valley of sorrow.

Thank You, Lord, for large, comfortable stepping stones.
For a while, they seemed small and nearly non-existent,
but faith and trust work wonders in enlarging the steps.

> FOR YE HAVE NOT PASSED THIS WAY HERETOFORE.
> *Joshua 3:4*

What new things can a widow possibly anticipate in a New Year (or anyone who has faced a major loss, when good anticipations become difficult)? Might she anticipate new expenses from new repair bills, new childhood accidents, newly broken items, new necessities, new doctor bills?

But who wants to dwell on the negative possibilities when God gives us positive possibilities with which to expand our minds?

» Psalm 40:3—*And he hath put a new song in my mouth, even praise unto our God: many shall see it, and fear, and shall trust in the Lord.* (There's going to be a new song of praise this year. Wait in anticipation, and then sing it with all your heart.)

» Isaiah 42:9—*Behold, the former things are come to pass, and new things do I declare: before they spring forth I tell you of them.* (God has faithfully fulfilled all things in the past, so why not trust Him and have faith in the new things to come?)

» Isaiah 65:16, 17—*That he who blesseth himself in the earth shall bless himself in the God of truth; and he that sweareth in the earth shall swear by the God of truth; because the former troubles are forgotten, and because they are hid from mine eyes. For, behold, I create new heavens and a new earth: and the former shall not be remembered, nor come into mind.* (Are we ready to forget the troubles of the past and expectantly await the coming of the new heavens and new earth?)

> *Anxiety is the rust of life, destroying its brightness*
> *and weakening its power. A childlike and abiding trust in*
> *God is its best preventive and remedy.*
>
> —Edwards

> FOR YE HAVE NOT PASSED THIS WAY HERETOFORE.
> *Joshua 3:4*

Here are a few more new things to anticipate in this New Year, and I'm sure God has many more new things for us. Let's just be alert to His goodness.

» Lamentations 3:22, 23—*It is of the Lord's mercies that we are not consumed, because his compassions fail not. They are new every morning: great is thy faithfulness.* (Let us every morning say, "Thank You for Your new mercies today, Lord.")

» Ephesians 4:24—*And that ye put on the new man, which after God is created in righteousness and true holiness.* (Ah, this is our daily responsibility, putting on a new man which every day becomes more like Christ.)

» Revelation 21:3-5—*Behold, the tabernacle of God is with men, and he will dwell with them, and they shall be his people, and God himself shall be with them, and be their God. And God shall wipe away all tears from their eyes; and there shall be no more death, neither sorrow, nor crying, neither shall there be any more pain: for the former things are passed away. And he that sat upon the throne said, Behold, I make all things new.* (What wonderful new things and new times we can expect. Imagine—no more tears, fears, or worries! No more death or dark valley times! No more disease, sickness, or pain! No more feeling abnormal, no more loneliness, no more adjustments to changes, no more decisions to make, no more battles with Satan. Our human minds find it difficult to think in heavenly terms!)

"What a day that will be, when the Savior we will see."

> **WHILE THE EARTH REMAINETH, SEEDTIME AND HARVEST,**
> and cold and heat, and summer and winter,
> and day and night shall not cease.
> *Genesis 8:22*

Our orderly God promises us orderly seasons; they will be in operation and succession as long as the earth remains. God's promises are sure; on them we can rest and depend.

This is an especially comforting thought to me. My grief is not orderly. I cannot depend on my emotions. This hurt feels like it will last forever! But, just as there are natural seasons, so there are seasons in life. This too will pass; another season will come. Thank God!

Perhaps you are in a winter storm of sickness or pain. Or, maybe you are having a blizzard of trials and troubles. Maybe you feel almost crushed under an avalanche of losses. You feel like winter will last forever.

When the winter season comes, observe the big, old apple tree. It is ugly, bent, bare, and seemingly useless.

Will it always look like that?

The winter season in life comes. It is ugly, bent, bare, and seemingly useless.

Will life always be so painful and unfocused, so ugly and fruitless?

> **WHILE THE EARTH REMAINETH, SEEDTIME AND HARVEST,**
> and cold and heat, and summer and winter,
> and day and night shall not cease.
> *Genesis 8:22*

God has promised us seasons in orderly operation. God is orderly and expresses himself in love and life.

Spring slowly seeps into winter, sending the ice and snow running to the streams and rivers. Soon, green sprigs of life appear everywhere. Little by little, barren, brown branches push out green buds. The ugly, old apple tree shows signs of life; the promise of something better appears. Shall we settle for those fresh, frilly green leaves only?

The springtime of life promises growth and vigor, an awakening to zest and zeal for living.

Am I content just with living the new life,
or does God have yet more to offer?

> WHILE THE EARTH REMAINETH, SEEDTIME AND HARVEST,
> and cold and heat, and summer and winter,
> and day and night shall not cease.
> *Genesis 8:22*

God has promised us the seasons in succession. God is in charge; He is a timely God. God brings changes—never too early or too late—if I cooperate. All nature simply flows with the mind of God. How beautiful!

After winter, comes spring. After spring, comes summer. Suddenly, the old apple tree appears young, dressed in frilly, pink blossoms. She never looked more alive—or did she?

The blossoms smell so fragrant and look so beautiful. In the summer of life we sometimes say, "It is enough to be beautiful and fragrant. Let me always keep my blossoms. Let me never see winter again."

But trees never just keep on blossoming. Life's summer is never everlasting.

Lo, another change is in progress.

> WHILE THE EARTH REMAINETH, SEEDTIME AND HARVEST,
> and cold and heat, and summer and winter,
> and day and night shall not cease.
> *Genesis 8:22*

God has promised us the seasons. As sure as day follows night, fall follows summer, and so the fruit follows the blossoms.

Fall produces the fruit, right? Wrong! Winter, spring, summer, and fall—it takes all the seasons to make the fruit. And it is the fruit that makes it an apple tree. It is the fruit that adds the value to the tree's life.

Always remember, what you see and feel in one season is only a part of the whole picture. Never judge life by one season. Never judge yourself or another person by the season you (he\she) are in. The essence of who we are can only be measured after all the seasons have been experienced.

If you quit in the winter, you'll never experience the feeling of new life in the spring, the beauty and fragrance of the summer, or the fruit of the fall. Never let the pain of one season destroy the joy and benefit of the other seasons. Live each season to the best of your ability by God's grace.

God bless you and guide you through all the seasons of your life.

This winter season of grief will give way to another spring, summer, and fruitful fall.

Reflections
of God's Grace in Grief

Small Size: 4 3/4" x 6 1/4"
Bechtel | 229 pages | Padded Hardcover | $12.99
Item #REF70332

Where do you go when your heart is breaking, when your life is crushed with loses that can never be replaced here on earth? Where do you go to find comfort and strength to make it through the lonely days and wearisome nights? Where do you find light to guide you through the dark valley of the shadow of death?

God and His Word supply the comfort, strength, courage, and light to keep on living when life makes no sense. This book is for those who have lost or for those who desire to relate to one dealing with grief. These meditations were written out of the author's own experience of three loses in five years and adjusting to a "new normal" of living.

ORDER FORM

To order, send this completed order form to:

Vision Publishers
P.O. Box 190 • Harrisonburg, VA 22803
Phone: 877.488.0901 • Fax: 540-437-1969
E-mail: orders@vision-publishers.com
www.vision-publishers.com

Name	Date
Mailing Address	Phone
City	State	Zip

Reflections - Revised Edition Qty. _____ x $17.99 ea. = _____
Reflections - Original Edition Qty. _____ x $12.99 ea. = _____

(Please call for quantity discounts - 877-488-0901)

Price _____
Virginia residents add 5% sales tax _____
Ohio residents add applicable sales tax _____
Shipping & handling - Add 10% of your total order plus $3.00
Grand Total _____

❏ Check #_____
❏ Money Order   ❏ Visa
❏ MasterCard   ❏ Discover

All Payments in US Dollars

Name on Card _____
Card # __|__|__| __|__|__|__| __|__|__|__| __|__|__|__|
3-digit code from signature panel __|__|__| Exp. Date __|__|__|__|

Thank you for your order!
For a complete listing of our books request our catalog.
Bookstore inquiries welcome